全国高职高专"十三五"贯穿式+立体化创新规划教材

电子商务网站建设企业案例

范兴昌　主　编

清华大学出版社
北　京

内 容 简 介

"电子商务网站建设"是高职院校电子商务专业的一门核心课程,它对于培养学生的商务网站建设与维护能力具有重要意义。

本书以一个基于 MVC 技术、采用面向切面的方式开发的一个电子商务系统的完整项目为案例,详细介绍了使用 Java Web 技术进行 Web 开发的基础知识和编程技巧、网站开发模式和开发过程,主要内容包括课程基础知识;搭建 Java Web 开发环境;Qzmall 电子商城系统设计;JQuery 和 JSTL 标签应用;商品显示模块设计;用户模块设计;购物车与订单模块设计;后台维护模块设计。本书将知识介绍和技能训练有机结合,融教、学、练、思于一体,适合案例教学、任务驱动、理实一体化的教学模式。

本书可作为高职高专电子商务专业的教材使用,也适合自学 Java Web 程序设计的读者使用。

本书封面贴有清华大学出版社防伪标签,无标签者不得销售。
版权所有,侵权必究。侵权举报电话: 010-62782989 13701121933

图书在版编目(CIP)数据

电子商务网站建设企业案例/范兴昌主编. —北京:清华大学出版社,2018
(全国高职高专"十三五"贯穿式+立体化创新规划教材)
ISBN 978-7-302-51121-2

Ⅰ. ①电… Ⅱ. ①范… Ⅲ. ①电子商务—网站建设—高等职业教育—教材 Ⅳ. ①F713.361.2 ②TP393.092

中国版本图书馆 CIP 数据核字(2018)第 201398 号

责任编辑:陈冬梅
装帧设计:杨玉兰
责任校对:李玉茹
责任印制:董 瑾

出版发行:清华大学出版社
网　　址:http://www.tup.com.cn, http://www.wqbook.com
地　　址:北京清华大学学研大厦 A 座　　邮　编:100084
社 总 机:010-62770175　　邮　购:010-62786544
投稿与读者服务:010-62776969, c-service@tup.tsinghua.edu.cn
质量反馈:010-62772015, zhiliang@tup.tsinghua.edu.cn
课件下载:http://www.tup.com.cn, 010-62791865

印 装 者:北京嘉实印刷有限公司
经　　销:全国新华书店
开　　本:185mm×260mm　　印 张:13.25　　字 数:322 千字
版　　次:2018 年 9 月第 1 版　　印 次:2018 年 9 月第 1 次印刷
定　　价:39.00 元

产品编号:077387-01

前　　言

"电子商务网站建设"是高职院校电子商务专业的一门核心课程，它对于培养学生的商务网站建设与维护能力具有重要意义。

本书依据网站建设与维护人员所应具备的职业能力，以真实的工程项目"Qzmall 电子商城"为工作任务，以 Java Web 为教学平台，以 Java 为程序设计语言，将"Qzmall 电子商城"的开发与维护分解为 7 个章节，每个章节充分体现相应的知识要点和结构。本书主要具有以下几个特点。

1. 准确的课程定位

根据电子商务企业对网站设计人员的基本要求，将课程目标定位为培养掌握 Java Web 基本开发技术的网站设计与维护人员，确保课程设置和课程内容对接职业标准和岗位要求。

2. 充分体现职业性的特点

采用 MVC 开发模式，注重对学生团队合作精神的培养，通过指导学生完成一系列实际工作任务来达到职业能力目标的要求，重点培养学生解决实际问题的能力。

3. 图文并茂，内容丰富

根据目前高职学生的认知能力和综合素质，在教材内容的处理上做到知识结构合理，图文并茂，内容丰富。在内容的安排上，立足高职学生理论基础知识较差，接受理论性知识的能力有限的现状，先易后难，使教材内容在够用、实用、会用等方面体现出特色，符合高职教育规律，让学生掌握必要的基本原理和基本操作技能。

4. 每个章节开头列出应掌握的知识能力目标

每个章节先介绍、分析完成部分模块的流程设计和相关知识点，再层层展开，按照项目开发的基本流程完成模块开发与设计，通过强化相关知识与技能的实际应用，从而提高学生的学习兴趣。每章后面安排了课后训练，习题侧重实际应用，注重培养学生的动手能力。

5. 有效整合教材内容与教学资源，打造立体化和自学学习式的新型教材

在编写教材的同时，还开发了涵盖课程标准、PPT、完整项目实例和课程开发所需要的相关软件及课程开发的项目原型和相关素材等在内的丰富的教学资源。

本书在讲解如何运用 Java Web 知识进行电子商务网站建设的同时，还涵盖了 jQuery、JSTL 标签方面的知识，对于这部分内容，我们只讲授和本课程有关的内容，超出本课程的内容，我们没有涉及。本书所有的源代码都在 Windows 7+JDK 1.8+Tomcat 8.0+

IntelliJ IDEA 2017 环境下调试通过，第四章的 jQuery 代码在 HBuilder 环境下调试通过。

　　参与本书编写的范兴昌等作者来自电子商务专业教学一线，具有丰富的教学经验，同时也了解企业对电子商务人才的专业需求。因此，本书既注重理论知识的系统性，又注重培养学生的动手和实践能力。

编　者

目　录

第一章　绪论 1
　　一、课程设计理念与思路 2
　　二、本课程的授课对象 2
　　三、课程的主要内容 2
　　四、教学设计 3
第二章　搭建网站开发环境 5
　第一节　搭建 Java Web 开发环境 6
　　一、Java Web 运行环境简介 6
　　二、JDK 的安装、配置与验证 7
　　三、安装与配置 Tomcat 9
　　四、搭建 MySQL 数据库环境 11
　第二节　利用 IntelliJ IDEA 创建 Java Web
　　　　　项目 12
　　一、IntelliJ IDEA 简介 12
　　二、利用 IntelliJ IDEA 实现热部署 16
　第三节　Java Web 程序概述 16
　　一、Java Web 程序的基本组成 16
　　二、Java Web 程序的目录结构 17
　　三、在 IDEA 里建立包、接口类、
　　　　接口实现类、servlet 18
　第四节　Servlet 基础知识 19
　　一、什么是 Servlet 19
　　二、Servlet 基本架构 21
　第五节　大型企业的网站开发模式 22
　　一、JSP Model1 22
　　二、JSP Model2 23
　　三、MVC 模式 24
　课后训练 25
第三章　Qzmall 电子商城系统设计 27
　第一节　Qzmall 电子商城系统概述 28
　　一、前台购物系统 28
　　二、后台功能需求 29
　第二节　设计并创建电子商城系统
　　　　　数据库 29
　　一、MySQL 外键设置方式 30
　　二、Qzmall 数据库表 30
　　三、设置触发器 33
　第三节　电子商城详细设计 34
　　一、实体类设计 34
　　二、定义数据库的连接类、
　　　　查询方法 36
　　三、创建 Web 应用过滤器 39
　课后训练 42
第四章　jQuery 和 JSTL 标签应用 45
　第一节　jQuery 简介 46
　　一、jQuery 语法 46
　　二、jQuery 选择器 47
　　三、jQuery 事件 47
　　四、jQuery 动画 49
　　五、jQuery 遍历 49
　第二节　轮播图效果实现 51
　　一、轮播图页面(CSS)的实现 51
　　二、手动单击左右按钮控制轮播的
　　　　切换 53
　　三、无缝轮播效果的实现 54
　第三节　Ajax 异步请求 57
　　一、Ajax 异步请求的概念 57
　　二、异步请求在用户注册中的应用 58
　第四节　JSTL 常用标签 62
　　一、EL 表达式介绍 62
　　二、JSTL 标签应用 62
　课后训练 66

第五章　商品显示模块设计 69

第一节　首页子模块设计 70
　　一、首页子模块功能分析 70
　　二、基础知识 70
　　三、首页子模块功能实现 71

第二节　商品分类显示子模块设计 81
　　一、商品分类显示子模块功能分析 ... 81
　　二、分页查询要求的基本数据 81
　　三、商品分类显示子模块功能实现 ... 81

第三节　商品详情子模块设计 92
　　一、商品详情显示功能分析 92
　　二、商品详情显示功能实现 92

课后训练 97

第六章　用户模块设计 99

第一节　用户注册子模块设计 100
　　一、用户注册的流程 100
　　二、基础知识 100
　　三、用户注册功能实现 101

第二节　用户登录子模块设计 107
　　一、用户登录基本流程 107
　　二、用户登录功能的实现 108

第三节　用户个人中心子模块设计 115
　　一、个人中心模块主要功能和
　　　　基本流程 115
　　二、用户个人信息维护功能实现 115

课后训练 122

第七章　购物车与订单模块设计 125

第一节　购物车模块设计 126
　　一、购物车的基本流程 126
　　二、购物车模块设计 126

第二节　生成订单子模块设计 137
　　一、生成订单模块的基本流程 137
　　二、生成订单子模块功能实现 137

第三节　我的订单子模块设计 145
　　一、我的订单子模块基本流程 145
　　二、我的订单子模块主要
　　　　功能实现 145

课后训练 162

第八章　后台维护模块设计 165

第一节　添加商品子模块设计 166
　　一、文件上传组件
　　　　commons-fileUpload 简介 166
　　二、添加商品功能实现 167

第二节　商品维护子模块设计 174
　　一、商品维护基本流程 174
　　二、商品维护主要功能实现 175

第三节　用户订单后台维护子模块
　　　　　设计 186
　　一、用户订单维护基本流程 186
　　二、用户订单后台维护功能实现 186

课后训练 199

习题答案 201

参考文献 203

第一章
绪　论

一、课程设计理念与思路

本课程设计遵循以职业能力培养为重点,基于工作过程进行课程开发与设计的理念,切实深化职业教育教学改革,提高课程建设水平和人才培养质量,体现了课程职业性、实践性和开放性的要求。"电子商务网站建设"课程涉及面广,相关知识多,为了科学合理选取教学内容和开展教学设计,笔者总结以往教学经验和教学效果,开展大量课程建设调研,采纳专家们提出的建议和意见,与企业共同开发课程,并参照了人力资源与社会保障部颁布的《电子商务师国家职业标准》。根据建设任务设计电子商务网站建设课程的教学内容,保障了课程内容选取的科学合理,并且具有针对性和适用性。课程设计充分考虑学生的认知水平和认知能力,把握学生的认知规律,合理地组织教学内容,按照由简单到复杂、由局部到整体、由易到难循序渐进地讲授知识和训练技能。

本课程强调以学生为中心的授课模式,学生主要在老师的引导下,对一个电商平台进行分析、设计与开发,按照软件开发的工作流程完成规划、设计、开发、测试等过程,提交网站建设计划书、工作日志及系统设计的源代码,让学生经历真实的网站开发过程,使学生走向工作岗位后,可以实现与工作岗位的无缝对接。

在教学过程中,采用团队分工的形式,每组最好不超过 5 人,要求分工明确,任务具体,团队协作融洽;提交的文档资料齐全、技术运用恰当。

二、本课程的授课对象

"电子商务网站建设"是电子商务专业必修课程之一,是一门实践性很强、面向应用的课程。它把基本概念和基本理论融入具体的网站建设与管理中。通过本课程的学习,要求能够理解网站建设的整体概念和基本步骤,掌握网站规划、设计、制作、维护的基本内容,同时能够熟练使用常用的工具软件进行网站的建设与维护。本课程授课对象主要是掌握了 JSP 程序设计基本理论和基本应用技能的电子商务专业的学生,也可为其他相关专业的学生从事网站开发提供借鉴和参考。

三、课程的主要内容

本书是作者在总结了多年教学经验的基础上编写的,全书围绕一个实际项目,从视图层、控制层、业务逻辑层 3 个层次全面、翔实地介绍了电子商务网站建设与开发的过程。全书共分为 8 章,第一章主要介绍本书的主要内容及设计思想;第二章介绍如何搭建网站开发环境,主要包括 Java Web 相关软件的安装及运行环境配置、MVC 开发模式及其优点;第三章介绍 Qzmall 电子商城系统设计,主要包括系统需求分析、数据库设计和系统详细设计;第四章介绍 jQuery 和 JSTL 标签应用,主要内容包括 jQuery 语法及应用、异步请求、JSTL 标签应用;第五章介绍商品显示模块设计,主要内容包括商品分类显示子模块设计、商品详情子模块设计;第六章介绍用户模块设计,主要内容包括用户注册子模块设计、用户登录子模块设计、用户个人中心子模块设计；第七章介绍购物车与订单模块设计，主要内容包括购物车模块设计、生成订单子模块设计和我的订单子模块设计；第八章介绍后台维护模块设计，主要内容包括商品维护子模块设计和用户订单后台维护子模块设计。

四、教学设计

(一)教学内容模块设计

"电子商务网站建设"是一门是以培养学生的企业网站建设能力为主要目标的课程，相关理论知识必须在技能训练过程中得以理解和掌握，职业态度和习惯要经过持续的训练得以养成。本书根据软件企业对基于 Web 开发能力的实际需求，坚持理论够用、适用、实用原则，以项目为中心，以能力为本位，将电子商务网站建设基本应用的知识和技能重新进行组合，形成了八大模块的教学内容。

(二)教学内容模块与项目功能模块的对应设计

教学内容模块与项目功能模块的对应如表 1-1 所示。

表 1-1 教学内容模块与项目功能模块的对应

序 号	教学内容模块	项目功能模块
1	商城需求分析、数据库设计与连接	网站系统设计
2	jQuery 应用	网站前台设计
3	JSTL 标签应用	商品模块设计
4	Servlet 读取表单数据和 session 数据	用户模块设计
5	应用过滤器进行身份验证	购物车与订单设计
6	文件上传工具应用	商品与订单后台维护

(三)实践环节的系统化设计

遵循"任务驱动、案例教学、理论实践一体化"的教学模式，通过精选真实项目，将项目精心分解，让学生在学习案例的同时，掌握 Java Web 开发技术，进而培养项目开发能力。同时，将理论教学和实践教学在同一教学时间和教学地点开展，将实践环节(课堂模拟、课堂实践、课外拓展、单元实践、综合实训)进行系统化设计，体现"学生为主体，教师为主导"的教学思想，实现"教、学、做"的完美统一。

在教学过程中，针对每一个教学单元，在实训室进行教学，授课时边讲边练，以调动学生学习的积极性和主动性，融教、学、做于一体，通过操作训练提高学生对课程技能点、知识点的理解和掌握。

本课程开发了丰富的数字化教学资源，如表 1-2 所示。

表 1-2　数字化教学资源

序号	资源名称	表现形式与内涵
1	课程标准	Word 电子文档，包含教学目标、教学目录、学时分配，可供教师备课使用
2	授课计划	是教师组织教学的实施计划表，包括具体教学进程、授课内容及时间、课外作业、授课方式等
3	PPT	PPT 电子文档，可以直接使用，也可以根据需要进行修改
4	项目库	课内教学使用、课外学生进行学习和训练使用的所有项目源代码

上述资源的开发，可以弥补单一纸质教材的不足，有利于教师利用现代教育技术手段完成教学任务；同时也提高了教材的适用性与普及性，特别是部分教学条件较弱或教学条件较强但学生接受能力较弱的学校，教师利用资源结合教材，可以更好地组织教学活动。

第二章

搭建网站开发环境

电子商务网站建设企业案例

知识能力目标

1. Java Web 运行环境。
2. 了解 IntelliJ IDEA 开发工具的主要特点。
3. 学会 JDK 的安装与配置。
4. 学会 Tomcat 的安装与配置。
5. 学会 MySQL 安装。
6. 学会 IntelliJ IDEA 开发工具的基本应用。

问题提示

Java Web 是用 Java 技术来解决相关 Web 互联网领域的技术总称。Web 包括 Web 服务器和 Web 客户端两部分。Java 在客户端的应用有 Java Applet；Java 在服务器端的应用非常丰富，如 Servlet、JSP 和第三方框架等。Java 技术为 Web 领域的发展注入了强大的动力，它以 Java 语言为基础，与 HTML 语言紧密结合，可以很好地实现 Web 页面设计和业务逻辑的分离，可以让 Web 程序员专注于业务逻辑的实现，大大提高了系统的执行性能。随着 Internet 的发展与普及，基于 Web 应用系统的开发也成为软件行业的主流，与 ASP 与 PHP 相比有着明显优势的 Java Web 开发技术，在 Web 开发中占据主导地位，如著名的淘宝、京东等电子商务网站都是用它来进行开发的。本书也是重点讲解如何运用 Java Web 技术进行电子商务网站的开发。

问题：利用 Java Web 技术进行电子商务网站开发，首先要做哪些准备并使用什么开发工具呢？

第一节　搭建 Java Web 开发环境

一、Java Web 运行环境简介

为了能够编写 Java Web 程序，至少需要具备两个基本条件：一是需要在计算机中安装 JDK，并进行相关的环境变量的设置；二是需要在计算机中安装 Java Web 引擎，即 Web 服务器，如 J2EE 服务器、IIS 服务器、Tomcat 服务器。基于 Java 的 Web 应用系统的开发和运行环境既包括客户端环境，也包括服务器环境，其中客户端运行环境只需要使用浏览器即可，推荐使用火狐浏览器(Mozilla Firefox)或谷歌浏览器(Google Chrome)，而服务器端的运行环境通常使用 Tomcat 来进行搭建。

搭建 Java Web 开发环境.flv

二、JDK 的安装、配置与验证

现阶段与 Java 相关的基础平台都是由 Oracle 公司提供的，开发人员可以通过 Oracle 公司的网站(http://www.oracle.com)了解有关 Java 的最新技术，并且可以下载相关的软件。

(一)安装 JDK

(1) 打开链接： http://www.oracle.com/technetwork/java/javase/downloads/jdk8-downloads-2133151.html，进入 JDK 1.8 下载官网，或者直接进入百度搜索 JDK 1.8，也可进入下载官网。下载完成后，双击 jdk_8.0.1310.11_64.exe，进入如图 2-1 所示的安装页面，选择 3 个可选功能，进行安装，如需更改安装目录，单击下边右侧的"更改"按钮即可。

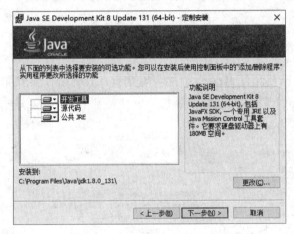

图 2-1　选择 JDK 安装选项

(2) 设置安装目录为：d:\Program Files\Java\jdk1.8.0_131\，如图 2-2 所示。

图 2-2　选择 JDK 安装目录

(3) JDK 安装过程中会出现两次安装提示：第一次是安装 JDK；第二次是安装 jre。建议两个都安装在同一个 Java 文件夹下的不同文件夹中(不能都安装在 Java 文件夹的根目录下，jdk 和 jre 安装在同一文件夹会出错)，如图 2-3 所示。

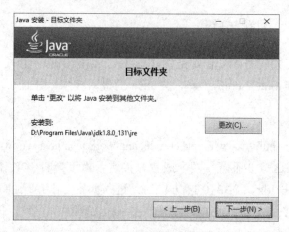

图 2-3　选择 jre 文件夹

(二)配置 JDK

安装完 JDK 后,需要配置 JDK 运行的环境变量,具体操作步骤如下。

(1) 右击"计算机"图标,选择"属性"→"高级系统设置"→"高级"→"环境变量"命令,如图 2-4 所示。

图 2-4　配置 JDK 环境变量

(2) 在"系统变量"对话框中单击"新建"按钮,设置"变量名"为 JAVA_HOME,"变量值"为 D:\Program Files\Java\jdk1.8.0_131,如图 2-5 所示。

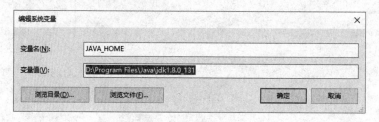

图 2-5　配置 JAVA_HOME 环境变量

(3) 编辑 Path 环境变量。

在"编辑环境变量"窗口中分别输入 %JAVA_HOME%\bin 和%JAVA_HOME%\jre\bin,

如图 2-6 所示。

图 2-6　编辑 Path 环境变量

(4) 编辑系统 CLASSPATH 变量，如图 2-7 所示。

变量值填写 ".;%JAVA_HOME%\lib;%JAVA_HOME%\lib\dt.jar;%JAVA_HOME%\lib\tools.jar"(注意最前面有一个点号)，系统变量配置完毕。

图 2-7　编辑 CLASSPATH 系统变量

(5) 检验是否配置成功，运行 cmd 并输入 java –version，若如图 2-8 所示显示版本信息，则说明安装和配置成功。

图 2-8　检验是否配置成功

三、安装与配置 Tomcat

Tomcat 服务器是当今使用最广泛的 Servlet/JSP 服务器，它运行稳定、性能可靠，是学习 JSP 技术和中小型企业应用的最佳选择。

Tomcat 的主页地址为 http://tomcat.apache.org，用户可以通过该网站的链接进入如图 2-9 所示的 Tomcat 下载页面，然后选择 64 位安装版本下载并安装即可。

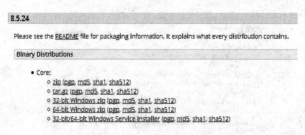

图 2-9　Tomcat 下载页面

Tomcat 安装路径是：C:\Program Files\Apache Software Foundation\Tomcat 8.0。

安装 Tomcat 后，在计算机→属性→高级→环境变量→系统变量中添加以下环境变量。

（1）CATALINA_HOME：C:\Program Files\Apache Software Foundation\Tomcat 8.0(TOMCAT 安装路径)。

（2）CATALINA_BASE：C:\Program Files\Apache Software Foundation\Tomcat 8.0。

（3）TOMCAT_HOME：C:\Program Files\Apache Software Foundation\Tomcat 8.0。

然后修改环境变量中的 CLASSPATH，把 Tomcat 安装目录下 common\lib 目录下的 servlet-api.jar 追加到 CLASSPATH 中去，修改后的 CLASSPATH 如下：

CLASSPATH=.;%JAVA_HOME%\lib\dt.jar;%JAVA_HOME%\lib\tools.jar;%CATALINA_HOME%\lib\servlet-api.jar。

接着可以启动 Tomcat，在 IE 浏览器中访问 http://localhost:8080，如果看到如图 2-10 所示的 Tomcat 欢迎页面，则说明安装成功了。

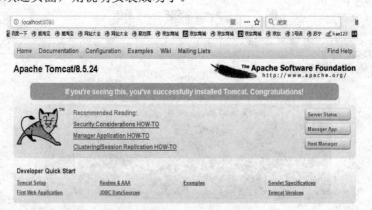

图 2-10　Tomcat 安装成功页面

小提示：Tomcat 有绿色版和安装版两种，绿色版下载之后直接解压到某一个目录，然后按照提示进行安装。

四、搭建 MySQL 数据库环境

(一)MySQL 安装

MySQL5.0 版本是当前比较稳定的安装版本。MySQL 的下载地址为 http://www.jb51.net/softs/2193.html。

下载完毕后,双击安装程序,按照操作提示,即可完成安装,安装步骤略。

(二)安装 MySQL 图形化管理工具——Navicat for MySQL

Navicat for MySQL 是一套专为 MySQL 设计的高性能数据库管理及开发工具。它可以用于任何版本为 3.21 或以上的 MySQL 数据库服务器,并支持大部分 MySQL 最新版本的功能,包括触发器、存储过程、函数、事件、视图、管理用户等,非常适合专业人员及入门新手使用,是一款非常易学的软件。软件界面设计直观简洁,兼容性强,支持各种编辑功能和数据管理。

课堂技能训练:

【操作内容】运用已下载好的 JDK、Tomcat、MySQL 等软件工具,配置 Java Web 开发环境。

【操作要求】
(1) 正确安装配置 JDK;
(2) 正确安装配置 Tomcat;
(3) 正确安装配置 MySQL 和 MySQL 图形化管理工具——Navicat for MySQL。

【知识拓展】ASP、PHP、JSP 三种网站开发技术比较

ASP 全名为 Active Server Pages,是一个 Web 服务器端的开发环境,利用它可以产生和执行动态的、互动的、高性能的 Web 服务应用程序。ASP 采用脚本语言 VBScript(Java Script)作为自己的开发语言。

PHP 是一种跨平台的服务器端的嵌入式脚本语言。它大量地借用 C 和 Perl 语言的语法,并耦合 PHP 自己的特性,使 Web 开发者能够快速地写出动态程序产生页面。它支持目前绝大多数数据库。PHP 是完全免费的,而且用户可以不受限制地获得源码,甚至可以在其中加进程序员自己需要的功能。

JSP(Java Server Pages)是 Sun 公司推出的一种动态网页技术。JSP 技术以 Java 语言作为脚本语言,熟悉 Java 语言的人可以很快上手。

JSP 本身虽然也是脚本语言,但是却和 PHP、ASP 有着本质的区别。PHP 和 ASP 都是由语言引擎解释执行程序代码,而 JSP 代码却被编译成 Servlet 并由 Java 虚拟机执行,这种编译操作仅在对 JSP 页面的第一次请求时发生。因此普遍认为 JSP 的执行效率比 PHP 和 ASP 都高。JSP 是一种服务器端的脚本语言,最大的好处就是开发效率较高,JSP 可以使用 JavaBean 和 Servlet 来执行应用程序所要求的更为复杂的处理。但是这种网站架构因为

其业务规则代码与页面代码混为一团，不利于维护，因此并不适合大型应用的要求，取而代之的是基于 MVC 的 Java Web 架构。

第二节　利用 IntelliJ IDEA 创建 Java Web 项目

一、IntelliJ IDEA 简介

IntelliJ IDEA，是 Java 语言开发的集成环境，IntelliJ IDEA 在业界被公认为是最好的 Java 开发工具之一，尤其在智能代码助手、代码自动提示、重构、J2EE 支持、各类版本工具(git、svn、github 等)、JUnit、CVS 整合、代码分析、创新 GUI 设计等方面的功能可以说是超常的。这里我们用的是 IntelliJ IDEA 2017.1.15 的 64 位版本。

创建 Java Web 项目.flv

(1) 创建一个项目。

打开 IDEA，进入如图 2-11 所示的界面。

(2) 选择 Create New Project，进入如图 2-12 所示的界面，勾选 Web Appliction(3.1)复选框，并确认已勾选 Create web.xml 复选框。

图 2-11　打开 IDEA 首页

图 2-12　勾选 Web Appliction(3.1)复选框

(3) 单击 Next 按钮，进入如图 2-13 所示的界面，输入项目名称 qzmall 和项目本地位置。

(4) 单击 Finish 按钮，完成项目创建，进入如图 2-14 所示的界面，在 web\WEB-INF 下创建两个文件夹：classes 和 lib。classes 用来存放编译后输出的 class 文件，lib 用于存放第三方 jar 包。

(5) 配置文件夹路径，选择 File → Project Structure 命令，进入如图 2-15 所示的界面，选择左边菜单项 Modules，在右边选项卡中选择 Paths，接着选中 Use module compile output path 单选按钮，将 Output path 设为刚刚创建的 classes 文件夹。

选择 Dependencies 选项卡，将 Module SDK 设置为 1.8，单击右边的"+"号，选择 1 Jars or Directories，选择刚刚创建的 lib 文件夹，如图 2-16 所示。

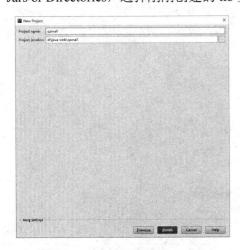

图 2-13　输入项目名称 qzmall 和
　　　　　项目本地位置

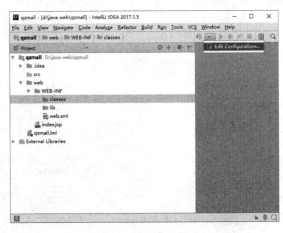

图 2-14　在 web\WEB-INF 下创建
　　　　　两个文件夹：classes 和 lib

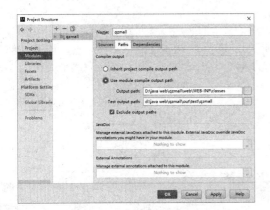

图 2-15　配置编译类文件夹

图 2-16　选择 lib 文件夹

紧接着选择 jar Directory 选项，如图 2-17 所示，完成项目文件夹配置。

图 2-17　配置 jar 包存放的文件夹

(6) 从网上下载 Tomcat 绿色版 8.5.8，把它放在 D:\Program Files 目录下，打开菜单 Run，选择 Edit Configuration，单击"+"号，展开左侧选项，选择 Tomcat Server→Local 选项，如图 2-18 所示。

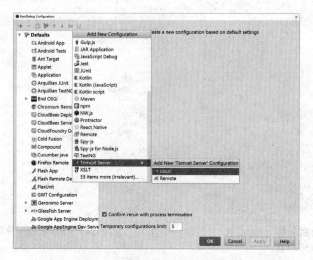

图 2-18　配置 Tomcat Server

(7) 配置 Tomcat 容器。

接上一步骤，单击 Local，进入如图 2-19 所示的界面，单击 Tomcat Server，展开右侧菜单，把 Tomcat 服务器名称更改为 qzmall。

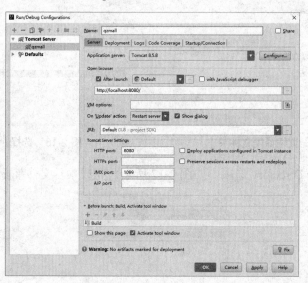

图 2-19　设置 Tomcat Server 名称

单击 Application server 右面的 Configure 按钮，弹出 Application Servers 窗口，单击 Tomcat Home 右边的选择按钮，选择本地安装的 Tomcat 目录，如图 2-20 所示，然后单击 OK 按钮。

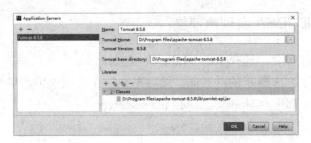

图 2-20　配置 Tomcat 安装目录

切换到 Deployment 选项卡，单击绿色"+"按钮，在旁边的 Application context 文本框中输入/qzmall，单击 Apply 按钮，如图 2-21 所示。

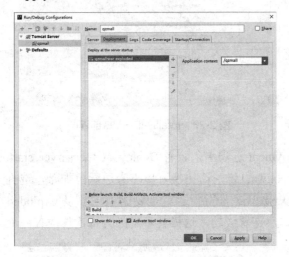

图 2-21　Deployment 选项卡

在 Run/Debug Configurations 对话框的 Server 选项卡中，设置 HTTP port 和 JMX port(默认值即可)，依次单击 Apply→OK 按钮，至此 Tomcat 配置完成。

(8) 编辑 index.jsp 文件。代码如下：

```
<%@ page contentType="text/html;charset=UTF-8" language="java" %>
<html>
  <head><title>第一个web项目</title></head>
  <body>
  Hello,World!第一个web项目运行成功!
  </body>
</html>
```

然后运行程序，显示如图 2-22 所示的页面。

图 2-22　第一个 Web 程序

二、利用 IntelliJ IDEA 实现热部署

(1) 打开 qzmall 项目，选择 Run→Edit Configurations 命令，单击 Tomcat Server，选择 qzmall，打开如图 2-23 所示的对话框，然后在 Server 选项卡中，将 On Update action、On frame deactivation 都选为 Update classes and resources。

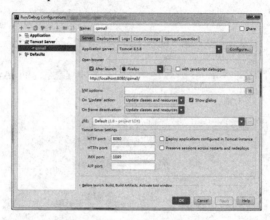

图 2-23　IntelliJ IDEA 热部署(一)

(2) 切换到 Deployment 选项卡，查看 Deploy at the server startup 框中 Tomcat 每次所运行的包，如果是 xxxx:war 包，请更换。单击旁边绿色加号，选择 xxxx:war exploded，然后单击红色减号将 xxxx:war 删除，也就是要让 IDEA 以 exploded 方式部署，如图 2-24 所示。单击 OK 按钮，热部署成功，这样就不用因为每次修改代码而重启了。

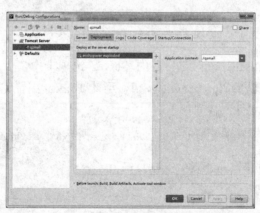

图 2-24　IntelliJ IDEA 热部署(二)

第三节　Java Web 程序概述

一、Java Web 程序的基本组成

一个 Java Web 应用程序是由一个或多个 Web 组件组成的集合。

JavaWeb 程序
概述.flv

这些 Web 组件一般被打包在一起，并在 Web 容器中运行。下面是一个典型 Java Web 应用程序的组成列表。

(1) Servlet。
(2) 在 Web 应用程序中使用的 Java 类。
(3) Java Server Pages(JSP)。
(4) JSP 标准标签(JSTL)和定制标签。
(5) 静态的文件，包括 HTML、图像、JavaScript 和 CSS 等。
(6) 描述 Web 应用程序的元信息(web.xml)。

二、Java Web 程序的目录结构

Java Web 程序的目录结构如图 2-25 所示。

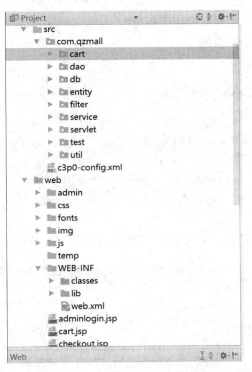

图 2-25 Java Web 程序的目录结构

(一)src 目录结构

src 目录下放的是各种 java 包。

(1) cart：购物车类包。
(2) dao：数据访问层类包。
(3) db：数据库。
(4) entity：与各个表相对应的实体类包。
(5) filter：过滤器类包。

(6) service：业务层类包。

(7) servlet：控制器包。

(8) test：测试包。

(9) util：工具包。

(10) c3p0-config.xml：数据库配置文件。

(二)web 目录结构

web 目录下放的都是与视图层有关的文件和目录。

(1) admin：放置后台操作的 jsp 文件和各种图片、css、js 文件。

(2) css：放置样式文件。

(3) fonts：放置字体和图标文件。

(4) img：放置图片文件。

(5) js：放置 js 文件。

(6) classes：放置编译文件。

(7) lib：放置各种 jar 包。

> 小提示：运用 web.xml 配置网站首页
> welcome-file-list 是一个配置在 web.xml 中的欢迎页，用于当用户在 url 中输入工程名称或者输入 Web 容器 url(如 http://localhost:8080/)时直接跳转到的页面。代码如下：
> ```
> <welcome-file-list>
>     <welcome-file>index.html</welcome-file>
>     <welcome-file>index.jsp</welcome-file>
> </welcome-file-list>
> ```

三、在 IDEA 里建立包、接口类、接口实现类、servlet

由于 IDEA 是 Java 语言开发的集成环境，所以建立包、接口类、接口实现类、servlet 相比其他软件就容易得多。

(一)在 IDEA 里建立包 ec

(1) 打开 IDEA，进入 qzmall 应用程序，右击 src 目录，选择 New→Package 命令，如图 2-26 所示。

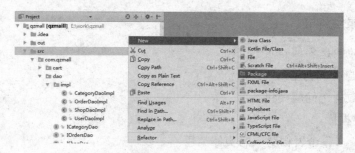

图 2-26　新建包

(2) 弹出如图 2-27 所示的对话框，输入包名 ec 即可。

(二)在 ec 包下新建接口类 IEcshop

右击 ec 包，选择 New→Java Class 命令，弹出如图 2-28 所示的对话框，输入 IEcshop，将 Kind 项设置为 Interface。

图 2-27　输入包名 ec

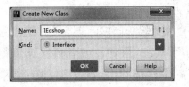

图 2-28　建立接口类 IEcshop

(三)在 ec 包下建立 Servlet→EcShopServlet

右击 ec 包，选择 New→Servlet 命令，输入名称 EcShopServlet 即可。

(四)在 ec 包下建立过滤器 Filter→EcShopFilter

右击 ec 包，选择 New→Filter 命令，输入名称 EcShopFilter 即可。

课堂技能训练：

【操作内容】建立 qzmall 应用程序，按照图 2-20 的要求，建立相应的包和文件夹，并进行热部署。

【操作要求】

(1) 建立 qzmall 应用程序。

(2) 在 web/WEB-INF 目录下创建两个文件夹：classes 和 lib，classes 用来存放编译后输出的 class 文件，lib 用于存放第三方 jar 包。

(3) 建立包、类、Servlet。

(4) 热部署。

第四节　Servlet 基础知识

一、什么是 Servlet

Java Servlet 是运行在 Web 服务器或应用服务器上的程序，它是作为来自 Web 浏览器或其他 HTTP 客户端的请求和 HTTP 服务器上的数据库或应用程序之间的中间层。使用 Servlet，可以收集来自网页表单的用户输入，呈现来自数据库或者其他源的记录，还可以动态创建网页。

(一)Servlet 的主要任务

(1) 读取客户端(浏览器)发送的显式的数据。包括网页上的 HTML 表单，也可以是来自 Applet 或自定义的 HTTP 客户端程序的表单。

(2) 读取客户端(浏览器)发送的隐式的 HTTP 请求数据。包括 Cookies、媒体类型和浏览器能理解的压缩格式等。

(3) 处理数据并生成结果。这个过程可能需要访问数据库，执行 RMI 或 CORBA 调用，调用 Web 服务，或者直接计算得出对应的响应。

(4) 发送显式的数据(即文档)到客户端(浏览器)。该文档的格式可以是多种多样的，包括文本文件(HTML 或 XML)、二进制文件(GIF 图像)、Excel 等。

(5) 发送隐式的 HTTP 响应到客户端(浏览器)。包括告诉浏览器或其他客户端被返回的文档类型(如 HTML)，设置 Cookies 和缓存参数，以及其他类似的任务。

(二)Servlet 的生命周期

Servlet 的生命周期可被定义为从创建直到毁灭的整个过程。以下是 Servlet 遵循的过程。

(1) Servlet 通过调用 init() 方法进行初始化。

(2) Servlet 调用 service()方法来处理客户端的请求，对每个请求均执行 service()方法。service() 方法由容器调用，service 方法在适当的时候调用 doGet、doPost、doPut、doDelete 等方法。因此，程序员不用对 service()方法做任何处理，只需要根据来自客户端的请求类型来重写 doGet() 或 doPost() 即可。

(3) Servlet 通过调用 destroy() 方法终止(结束)。

(三)Servlet 三大作用域

在 Java Web 开发中，Servlet 有三大域对象的应用：request、session、application。

1. request

request 表示一个请求，只要发出一个请求就会创建一个 request，它的作用域仅在当前请求中有效。

用处：常用于服务器间同一请求不同页面之间的参数传递，以及表单的控件值传递。

2. session

服务器会为每个会话创建一个 session 对象，所以 session 中的数据可供当前会话中所有 Servlet 共享。

从用户打开浏览器会话开始，直到关闭浏览器会话才会结束。一次会话期间只会创建一个 session 对象。

用处：常用于 Web 开发中的登录验证界面(当用户登录成功后浏览器分配其一个 session 键值对)。

session 是服务器端对象，保存在服务器端。并且服务器可以将创建 session 后产生的 sessionid 通过一个 cookie 返回给客户端，以便下次验证。

3. Application(ServletContext)

作用范围：所有的用户都可以取得此信息，此信息在整个服务器上被保留。Application 属性范围值只要设置一次，则所有的网页窗口都可以取得数据。ServletContext 在服务器启动时创建，在服务器关闭时销毁，一个 Java Web 应用只创建一个 ServletContext 对象，所有的客户端在访问服务器时都共享同一个 ServletContext 对象。ServletContext 对象一般用于在多个客户端间共享数据时使用。

获取 Application 对象的方法(Servlet 中)：

```
ServletContext app01 = this.getServletContext();
app01.setAttribute("name", "kaixuan");    //设置一个值进去
```

服务器只会创建一个 ServletContext 对象，所以 app01 就是 app02，通过 app01 设置的值当然可以通过 app02 获取。代码如下：

```
ServletContext app02 = this.getServletContext();
app02.getAttribute("name");  //获取键值对
```

二、Servlet 基本架构

Servlet 基本架构的代码如下：

```
@WebServlet(name = "ClassNameServlet",urlPatterns = "/class.do")
public class ClassNameServlet extends HttpServlet {
    public void init() throws ServletException {
        //初始化代码...
    }
    protected void doPost(HttpServletRequest req, HttpServletResponse resp) throws ServletException, IOException {
        //Servlet 代码
    }
    protected void doGet(HttpServletRequest req, HttpServletResponse resp) throws ServletException, IOException {
        //Servlet 代码
    }
    public void destroy() {
        //终止化代码...
    }
}
```

【代码说明】

(1) @WebServlet(name = "ClassNameServlet",urlPatterns = "/class.do")是采用注解方式配置 Servlet，要求版本必须是 Servlet 3.0 以上版本。多数 Servlet 在实际应用中，init()和 destroy()一般不用写在里面。

（2）GET 请求来自一个 URL 的正常请求，或者来自一个未指定 METHOD 的 HTML 表单，它由 doGet()方法处理。POST 请求来自一个特别指定了 METHOD 为 POST 的 HTML 表单，它由 doPost()方法处理。

> 小提示：get 提交和 post 提交有何区别
> （1）get 一般用于从服务器上获取数据；post 一般用于向服务器传送数据。
> （2）请求的时候参数的位置有区别。get 的参数是拼接在 url 后面，用户在浏览器地址栏可以看到；post 的参数是放在 http 包的包体中。
> 比如用户注册，不能把用户提交的注册信息用 get 的方式，否则就相当于把用户的注册信息都显示在 url 上了，是不安全的。
> （3）能提交的数据有区别。get 方式能提交的数据只能是文本，且大小不超过 1024 字节；而 post 不仅可以提交文本，还可以提交二进制文件。
> 所以，上传文件就需要使用 post 请求方式。
> （4）Servlet 在处理请求的时候分别使用 doGet 和 doPost 方式进行处理请求。

第五节 大型企业的网站开发模式

一、JSP Model1

(一)传统的 JSP Model1

JSP 是独立的，自主完成所有的任务，如图 2-29 所示。

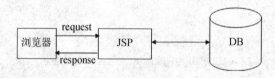

图 2-29 传统的 JSP Model1

(二)改进的 JSP Model1

JSP 页面与 JavaBeans 共同协作完成任务，如图 2-30 所示。

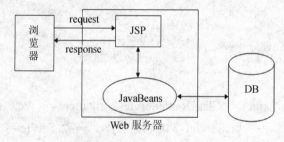

图 2-30 改进的 JSP Model1

(三)JSP Model1 的优点

这种架构模型非常适合开发业务逻辑不太复杂的小型 Web 项目，而且对 Java Web 开发人员的技术水平要求不高。在这种模式下，JavaBeans 用于封装业务数据，JSP 既负责处理用户请求，又显示数据。

(四)JSP Model1 的缺点

1. HTML 和 Java 强耦合在一起，导致页面设计与逻辑处理无法分离

大量使用这种模式，常会导致在 JSP 页面中嵌入大量的 Java 代码，当需要处理的商业逻辑非常复杂时，这种情况就会变得很糟糕。

2. 可读性差，调试困难，不利于维护

大量的 Java 代码使得 JSP 页面变得非常臃肿，可读性差。前端的页面设计人员稍有不慎，就有可能破坏关系到商业逻辑的代码。这造成了代码开发和维护的困难。

3. 功能划分不清

这种情况若在大型项目中同时出现，会功能划分不清，导致项目管理的困难。因此，这种模式只适用于中小规模的项目。

二、JSP Model2

(一)JSP Model2 中使用了三种技术：JSP、Servlet 和 JavaBeans

(1) JSP 负责生成动态网页，只用作显示页面。
(2) Servlet 负责流程控制，用来处理各种请求的分派。
(3) JavaBeans 负责业务逻辑和对数据库的操作。

(二)使用 JSP Model2 的交互过程

用户通过浏览器向 Web 应用中的 Servlet 发送请求，Servlet 接收到请求后实例化 JavaBeans 对象，调用 JavaBeans 对象的方法，JavaBeans 对象返回从数据库中读取的数据。Servlet 选择合适的 JSP，并且把从数据库中读取的数据通过这个 JSP 进行显示，最后 JSP 页面把最终的结果返回给浏览器，如图 2-31 所示。

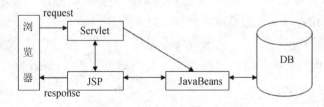

图 2-31　JSP Model2 的交互过程

(三)JSP Model2 的优点

(1) 消除了 JSP Model1 的缺点。
(2) 该模式适合多人合作开发大型的 Web 项目。
(3) 各司其职，互不干涉。
(4) 有利于开发中的分工。
(5) 有利于组件的重用。

(四)JSP Model2 的缺点

Web 项目的开发难度加大，同时对开发人员的技术要求也提高了。

三、MVC 模式

(一)MVC 模式的概念

MVC(Model-View-Controller)模式通过 JSP 技术来表现页面，通过 Servlet 技术来完成大量的事务处理工作，实现用户的商业逻辑。

在这种模式中，Servlet 用来处理请求的事务，充当了控制器(Controller)的角色，它负责响应客户对业务逻辑的请求并根据用户的请求行为，决定将哪个 JSP 页面发送给客户。JSP 页面处于表现层，也就是视图(View)的角色。JavaBeans 则负责数据的处理，也就是模型(Model)的角色。

Servlet+JSP+JavaBeans(MVC)模式适合开发复杂的 Web 应用。在这种模式下，Servlet 负责处理用户请求；JSP 负责数据显示；JavaBeans 负责封装数据。Servlet+JSP+JavaBeans 模式的程序各个模块之间层次清晰，Web 开发推荐采用此种模式。

(二)MVC 模式的优点

大部分用语言如 ASP、PHP 开发出来的 Web 应用，初始的开发模板就是混合层的数据编程。例如，直接向数据库发送请求并用 HTML 显示，开发速度往往比较快，但由于数据页面的分离不是很直接，因而很难体现出业务模型的样子或者模型的重用性。产品设计弹性力度很小，很难满足用户的变化性需求。MVC 要求对应用分层，虽然要处理额外的工作，但产品的结构清晰，产品的应用通过模型可以得到更好的体现。

最重要的是应该有多个视图对应一个模型的能力。在目前用户需求快速变化的情况下，可能有多种方式访问应用的要求。例如，订单模型可能有本系统的订单，也有网上订单，或者其他系统的订单，但对于订单的处理都是一样的。按 MVC 设计模式，一个订单模型以及多个视图即可解决问题。这样减少了代码的复制和代码的维护量，一旦模型发生改变，也易于维护。

课后训练

一、选择题

1. 关于 Eclipse 与 IDEA 的不同，下列选项中正确的是(   )。

   A. IDEA 可管理多个项目

   B. IDEA 进行专一的项目管理

   C. Eclipse 是一个开放的环境，可进行多项目管理

   D. Eclipse 是精准的单一项目管理

2. 在使用 IDEA 开发工具时，如果需要将代码结构补全，需要使用(   )组合键。

   A. Ctrl +Shift+Space   B. Ctrl+Alt

   C. Ctrl +Shift+Alt   D. Ctrl +Shift+Enter

3. 在 IDEA 中遇到项目乱码问题，应该怎么处理？(   )

   A. 将乱码代码或者注释去掉   B. 改变单个文件的编码方式

   C. 不予理会   D. 改变整个项目环境的编码方式

4. 在 IDEA 创建中，Java 项目的源程序存在于哪个文件夹下？(   )

   A. src   B. idea   C. out   D. WEB-INF

5. 在 Java Web 应用开发中，如果客户端的每次请求均调用 Servlet，则每次调用都会执行 Servlet 生命周期中的(   )方法。

   A. init()   B. destroy()   C. service()   D. close()

6. 在 JavaEE 中，初始化 Servlet 实例的时候，init()方法是(   )。

   A. 由程序员编写代码来调用执行

   B. 每次执行当前的 Servlet 时，由系统自动执行

   C. 当第一次执行当前的 Servlet 时，由系统自动执行

   D. 以上说法都不对

7. 在 Java Servlet API 中，HttpServletRequest 接口的(   )方法用于返回与当前请求相关联的会话，如果没有，则返回 null。

   A. getSession()   B. getSession(true)

   C. getSession(false)   D. getSession(null)

8. 关于会话属性，以下说法正确的是(   )。

   A. HttpSession.getAttribute(String)的返回类型是 Object

   B. HttpSession.getAttribute(String)的返回类型是 String

   C. 在一个 HttpSession()上调用 setAttribute("key", "value")时，如果这个会话中对应属性 key 已经有一个值，就会导致一个异常

   D. 在一个 HttpSession()上调用 setAttribute("key", "value")时，如果这个会话中对应属性 key 已经有一个值，则会导致这个属性原先的值被 String value 替换

9. 如果做动态网站的开发，以下( )可以作为服务器端脚本语言。
   A. HTML   B. JSP   C. JavaScript   D. Java
10. 在运行 Web 项目时，IE 提示"404 错误"，可能的原因包括( )。
   A. 未启动 Tomcat 服务   B. 未部署 Web 项目
   C. URL 中的上下文路径书写错误   D. URL 中的文件名书写错误
11. 在 MVC 模式体系结构中，( )是实现控制器的首选方案。
   A. JavaBeans   B. Servlet   C. JSP   D. HTML
12. 编写 Servlet 的 doPost 方法时，需要抛出的异常为( )。
   A. ServletException，IOException   B. ServletException，RemoteException
   C. HttpServletException，IOException   D. HttpServletException，RemoteException
13. 使用 MVC 模式设计的 Web 应用程序具有以下优点，除了( )。
   A. 可维护性强   B. 可扩展性强   C. 代码重复较少   D. 大大减少代码量
14. 在 MVC 模式中，( )层专用于客户端应用程序的图形数据表示，与实际数据处理无关。
   A. 模型   B. 视图   C. 控制器   D. 数据

二、实际操作题

配置 Java Web 开发环境，建立应用程序并建立相应的包和文件夹，最后进行热部署。

第二章 构建网站开发环境.pptx   第二章 习题答案.docx

第三章
Qzmall 电子商城系统设计

知识能力目标

1. 了解 Qzmall 电子商城系统基本设计思想。
2. 掌握数据库设计，掌握主键、外键的含义以及触发器的作用。
3. 理解过滤器的作用。
4. 学会数据库设计和表设计，学会设置主键、外键，学会触发器的设置。
5. 学会配置数据库连接。
6. 学会配置过滤器。

问题提示

建设网上商城之前，首先要了解用户的需求，确定商城的基本功能，需要通过数据库来实现和用户进行交互的复杂的商品显示、查找、登录、购物等功能。

问题：建立一个网站商城需要有哪些表？表与表之间的关系是什么？

第一节 Qzmall 电子商城系统概述

一、前台购物系统

(一)商品展示/搜索

(1) 商品查询。根据查询条件查询所有符合条件的商品信息。查询商品流程：用户输入查询条件并提交查询；系统根据用户提交的查询条件从数据库中查询商品并输出。

(2) 商品展示。首页显示本网站的最新商品。

(3) 商品浏览。输入需要查看信息的种类，查看该类型的所有产品。

(4) 商品详情。可以看到商品的详细信息，如商品名称、图片、价格、基本参数。

(二)购物车/订单

(1) 加入购物车。注册用户可以在浏览商品的过程中，把需要购买的商品放置在一起付款，单击商品下方的"加入购物车"按钮即可把商品加入购物车。

(2) 购物车查询。用户在购物过程中可以随时查看购物车中的商品，以了解所选购的商品信息。

(3) 注册用户选购商品后，在确认购买之前，可以对购物车的商品进行二次选择，既可以删除购物车中不要的商品，也可以修改购物车中的商品购买数量。

(4) 在用户确认购买后，系统会为注册用户生成购物订单，注册用户可以查看自己的订单信息以了解付款信息和商品配送情况。

(5) 查询、修改订单状态。用户登录成功后，可以修改订单状态，如付款、收货等。

(6) 取消订单。用户在未付款之前，可以取消订单。

(7) 订单查询。用户登录后，可以浏览或查询本人订单。

(三)会员

(1) 注册。系统考虑到用户购买信息的真实性，规定访客只能在系统中查看商品信息，不能进行商品的订购。但是访客可以进行注册，登记相关基本信息，成为系统的注册用户。

(2) 登录。用户输入用户名及用户密码可以进行登录。

(3) 信息修改。用户登录之后修改自己的账号、密码及其他个人信息。

二、后台功能需求

(一)商品分类维护

管理员进入后台系统后可以对商品分类信息进行增、删、改、查。

(二)商品维护

管理员进入系统后可以添加商品信息、修改已有商品信息、查询商品信息和查询某类及某件商品信息。

(三)订单

订单是用户在前台购物过程中生成的，后台管理员可以对已付款订单进行订单状态修改，同时根据订单情况通知物流配送人员进行商品配送。

(四)商品库存维护

商品库存会随着订单的完成而减少，管理员可以为缺货商品添加库存，并能查看单个商品的库存周转记录。

课堂技能训练：

【实训操作内容】登录淘宝、京东、苏宁等购物网站，分析这些网站主要有哪些功能，并比较它们功能的异同。

【实训操作要求】必须写出每个网站相比于其他电子商务网站有哪些特色。

第二节　设计并创建电子商城系统数据库

本系统采用数据库管理系统 MySQL 管理商品、分类、会员、订单、购物车等数据。数据库表的结构设计是应用软件中的关键部分，数据库表设计采用规范化关系模式的方法。在关系数据库中，规范化关系模式就是要解决关系模式中存在的插入异常、删除异常、修改复杂、数据冗余等问题。规范化的基本思想是消除数据依赖中不合适的部分，使模式中的各关系模式达到某种程度的"分离"，让一个关系描述一个概念、一个实体或者实体间的一种联

设计并创建 qzmall 数据库.flv

系，若多于一个概念就把它分离出去。

一、MySQL 外键设置方式

(一)cascade 方式

在父表上 update/delete(更新或删除)记录时，同步 update/delete 子表的匹配记录。

(二)set null 方式

在父表上 update/delete 记录时，将子表上匹配记录的列设为 null。
要注意子表的外键列不能为 not null。

(三)no action 方式

如果子表中有匹配的记录，则不允许对父表对应候选键进行 update/delete 操作。

(四)restrict 方式

同 no action，都是立即检查外键约束。

(五)set default 方式

父表有变更时，子表将外键列设置成一个默认的值，但 InnoDB 不支持此方式。

二、Qzmall 数据库表

根据系统功能描述和实际业务分析，进行 Qzmall 电子商城系统的设计，其主要数据及其内容如表 3-1～表 3-10 所示。

表 3-1 商品大类表(category)

表序号	1	表名	category	
含义	存储商品大类信息	关联表	subcatE. shop	
字段名	字段类型	字段长度	键别	描述
cid	int(自动增长)	11	主键	大类编号
cname	varchar	30		大类名称

表 3-2 商品子类表(subcate)

表序号	2	表名	subcate	
含义	存储商品子类信息	关联表	shop	
字段名	字段类型	字段长度	键别	描述
sid	int(自动增长)	11	主键	大类编号
cid	int	11	外键(category)	大类编号
sname	varchar	30		小类名称

表 3-3 商品信息表(shop)

表序号	3		表名	shop	
含义	存储商品信息		关联表	Category subcate	
字段名	类型	长度		键别	说明
pid	int(自动增长)	11		主键	商品编号
cid	int	11		外键(category)	大类编号
sid	int	11		外键(subcate)	小类编号
shopname	varchar	40			商品名称
shopinfo	varchar	100			基本信息
price	float	8,1			价格
stock	int	11			库存
shopdate	varchar	20			上架时间
image1	varchar	100			图片1
image2	varchar	100			图片2
image3	varchar	100			图片3
description	varchar	2000			说明

表 3-4 订单主表(orders)

表序号	4		表名	orders	
含义	存储订单主要信息		关联表	user	
字段名	类型	长度		键别	说明
orderId	varchar	20		主键	订单编号
username	varchar	30		外键(user)	会员名
truename	varchar	40			真实姓名
postcode	varchar	6			邮编
addtime	varchar	20			时间
phone	varchar	11			手机
address	varchar	100			数量
sum	float	8,1			总价
state	int	1			订单状态

表 3-5 订单明细表(ordersitem)

表序号	5		表名	ordersitem	
含义	存储订单明细信息		关联表	Orders shop	
字段名	类型	长度		键别	说明
oid	int(自动增长)	11		主键	编号
orderId	varchar	20		外键(orders)	订单编号

续表

字段名	类型	长度	键别	说明
pid	int	11	外键（shop）	商品编号
shopname	varchar	40		商品名称
price	float	100		售价
shopnum	int	11		数量

表 3-6 用户信息表(user)

表序号	6		表 名	user
含义	存储用户明细信息		关联表	
字段名	类型	长度	键别	说明
username	varchar	30	主键	编号
password	varchar	50		支付方式
truename	varchar	20		真实姓名
postcode	varchar	6		邮编
adddate	varchar	20		时间
phone	varchar	11		手机
address	varchar	100		数量
point	int	11		积分
question	varchar	10		密保问题
birthday	varchar	10		生日
answer	varchar	30		答案

表 3-7 广告表 (ad)

表序号	7		表名	ad
含义	存储广告		关联表	subcate
字段名	字段类型	字段长度	键别	描述
aid	int(自动增长)	11	主键	广告编号
sid	int	11	外键	大类编号
aimage	varchar	30		广告图片
atime	bigint	20		上架时间

表 3-8 会员积分表(point)

表序号	8		表名	point
含义	存储会员积分记录		关联表	user
字段名	字段类型	字段长度	键别	描述
pointid	int(自动增长)	11	主键	编号
username	varchar	30	外键	会员名
orderId	varchar	20	外键	订单编号
pointtime	bigint	20		积分时间
description	varchar	100		积分说明

表 3-9　商品库存记录表(stockitem)

表序号	9	表名	stockitem	
含义	存储商品库存记录	关联表	shop	
字段名	字段类型	字段长度	键别	描述
stockid	int(自动增长)	11	主键	库存编号
pid	int	11	外键	商品编号
num	int	11		数量
price	float	8,1		价格
stocktime	bigint	20		时间

表 3-10　管理员(admin)

表序号	10	表名	admin	
含义	存储商品大类信息	关联表		
字段名	字段类型	字段长度	键别	描述
aname	varchar	20	主键	管理员信息
apass	varchar	30		密码

三、设置触发器

```
表:point
类型：插入<!---insert--->描述本表插入记录的同时，更新user表的point字段值
begin
declare c int;
set c = (select point from user where username=new.username);
update user set point = c + new.mypoint where username = new.username;
end
类型：删除<!---delete--->描述本表删除记录的同时，更新user表的point字段值
begin
declare c int;
set c = (select point from user where username=old.username);
update user set point = c - old.mypoint where username = old.username;
end
表 : stockitem
触发器名称：adstockItem
类型：插入<!---insert--->描述插入记录的同时，商品表对应商品的库存应增加同样的数量
begin
declare c int;
set c = (select stock from shop where pid=new.pid);
update shop set stock = c+ new.shopnum where pid = new.pid;
end
触发器名称：destockItem
类型：删除<!---delete---> 描述删除记录的同时，商品表对应商品的库存应减少同样的数量
begin
declare c int;
```

```
set c = (select stock from shop where pid=old.pid);
update shop set stock = c-old.shopnum where pid = old.pid;
End
```

小提示：利用 navcat for mysql 添加触发器

打开 Qzmall 数据库后，右击表 point，单击"触发器"标签，再单击"添加触发器"按钮，输入触发器名称，选择触发方式，然后在"定义"文本框中输入触发器内容即可，如图 3-1 所示。

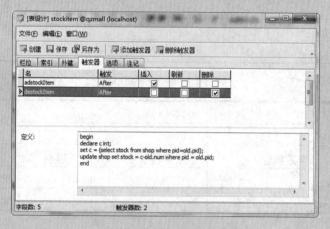

图 3-1 添加触发器

课堂技能训练：

【实训操作内容】利用 navcat for mysql 数据库操作可视化工具，建立 Qzmall 数据库。

【实训操作要求】

(1) 设置每个表的主键、字段类型及其长度。

(2) 建立表与表之间的关系(主键、外键约束)。

(3) 建立触发器(为 stockitem 表建立触发器，即在增加商品库存记录时，shop 表中库存数量能够一并更新)。

第三节　电子商城详细设计

一、实体类设计

实体类设计.flv

编写 Qzmall 商城实体类时应注意以下问题。

(1) 实体类的名字尽量和数据库表的名字对应相同。

(2) 实体类应实现 java.io.Serializable 接口。

(3) 实体类一般有无参和在参的构造方法。

(4) 实体类有属性和方法，属性对应数据库中表的字段，主要有 getter 和 setter 方法。

(5) 属性一般是 private 类型，方法一般是 public 类型，对于数据库自动生成的 ID 字段对应的属性的 set 方法应为 private。

实体类放在 com.qzmall.entity 包下，各个实体类的代码如下：

```java
//大类表(category)对应的实体类
public class Category {
    private int cid;//大类编号
    private String cname;//大类名称
    private ArrayList<SubCate> subCates = new ArrayList<SubCate>(0);
    //和子类表建立对应关系
    private ArrayList<Shop> shopArrayList=new ArrayList<Shop>(0);
    //和商品表建立对应关系
//set、get方法略 }
//子类表(subcate)对应的实体类
public class SubCate {
    private int sid;//子类编号
    private Category category;   //商品大类
    private String sname;        //小类名称
    private Set shops = new HashSet(0);//和商品表建立对应关系
//set、get方法略 }
//商品表(shop)对应的实体类
public class Shop {
    public Shop() {    }
    private int pid; //商品I
    private Category category;    //商品大类
    private SubCate subCate;      //商品子类
    private String shopname;      //商品名称
    private String shopinfo;      //商品信息
    private Float price;          //价格
    private int stock;            //库存
    private String shopdate;      //上架日期
    private String image1;        //图片1
    private String image2;        //图片2
    private String image3;        //图片3
    private String description;   //详细信息
//set、get方法略 }
//用户表(user)对应的实体类
public class User {
    public User() {    }
    private String username;      //登录账号
    private String password;      //登录密码(MD5加密)
    private String truename;      //真实姓名
    private String birthday;      //出生日期
    private String adddate;       //注册时间
    private int question;         //请输入密码问题
    private String answer;        //密码问题答案
    private String phone;         //手机
    private String postcode;      //邮编
    private String address;       //地址
    private int point;            //积分
//set、get方法略}
```

```java
//订单主表(orders)对应的实体类
public class Orders {
    private String orderId;          //订单编号
    private String username;         //订单用户名
    private String truename;         //真实姓名
    private String address;          //地址
    private String phone;            //手机号码
    private String postcode;         //邮编
    private String addtime;          //订单时间
    private float sum;               //金额
    private int state;               //订单状态
//set、get 方法略 }
//订单明细表(ordersitem)对应的实体类
public class OrdersItem {
    private int oid;                 //订单明细编号
    private String orderId;          //订单编号
    private int pid;                 //商品编号
    private String shopname;         //商品名称
    private float price;             //商品价格
    private int shopnum;             //商品数量
//set、get 方法略 }
```

小提示： 利用 IntelliJ IDEA 在实体类中插入 set、get 方法非常方便

右击类的空白处，选择 Generate 选项，然后出现 Generate 面板，选择 Getter 和 Setter 选项，如图 3-2 所示，选择想要生成 get、set 方法的属性，单击 OK 按钮，即可自动生成 get 和 set 方法。

图 3-2 为实体类属性添加 set 和 get 方法

二、定义数据库的连接类、查询方法

下面介绍如何定义数据库的连接类、查询方法。代码如下：

```java
//数据库连接管理类 ConnectionManager.java
package com.qzmall.db;
import java.sql.Connection;
```

数据库连接类
设计.flv

```java
import java.sql.SQLException;
import com.mchange.v2.c3p0.ComboPooledDataSource;
import com.mchange.v2.c3p0.DataSources;
public class ConnectionManager {
    private static ConnectionManager instance;
    //C3P0 的连接池对象
    private ComboPooledDataSource ds;
    private ConnectionManager() throws Exception {
        ds = new ComboPooledDataSource("mysql");
    }
    //单例模式获取数据库连接对象
    public static final ConnectionManager getInstance() {
        if (instance == null) {
            try {
                instance = new ConnectionManager();
            } catch (Exception e) {
                e.printStackTrace();
            }
        }
        return instance;
    }
//为了线程安全,同步
    public synchronized final Connection getConnection() {
        try {
            return ds.getConnection();
        } catch (SQLException e) {
            e.printStackTrace();
        }
        return null;
    }
    @Override
    protected void finalize() throws Throwable {
        DataSources.destroy(ds);   //关闭 datasource
        super.finalize();
    }
}
//数据库操作类 DBUtil.java
package com.qzmall.db;
import java.sql.Connection;
import java.sql.PreparedStatement;
import java.sql.ResultSet;
import java.sql.SQLException;
public class DBUtil {
    /*从 C3P0 数据库连接池中获取数据库连接对象*/
    public static Connection getConnection() {
        Connection conn = null;
        try {
            //从数据连接池中获取数据库连接对象
            conn = ConnectionManager.getInstance().getConnection();
        } catch (Exception e) {
```

```java
            e.printStackTrace();
        }
        return conn;
    }
    //释放资源
    public static void close(Connection connection,
        PreparedStatement psmtStatement, ResultSet resultSet) {
        try {
            if (resultSet != null) {
                resultSet.close();
            }
            if (psmtStatement != null) {
                psmtStatement.close();
            }
            if (connection != null) {
                connection.close();
            }
        } catch (Exception e) {
            e.printStackTrace();
        }
    }
//增、删、改的通用方法
    public static int executeUpdate(String sql, Object... objects) {
        int result = 0;
        Connection conn = null;
        PreparedStatement psmt = null;
        try {
            conn = getConnection();
            psmt = conn.prepareStatement(sql);
            if (objects != null) {
                for (int i = 0; i < objects.length; i++)
{psmt.setObject(i + 1, objects[i]);  //下标从1开始的
                }
            }
            result = psmt.executeUpdate();
        } catch (Exception e) {
            e.printStackTrace();
        } finally {
            close(conn, psmt, null);
        }
        return result;
    }
    //查询通用的方法
    public static Object executeQuery(String sql, IResultSetUtil rsHandler,
Object... objects) { Connection connection = null;
        PreparedStatement preparedStatement = null;
        ResultSet resultSet = null;
        try {
            connection = getConnection();
            preparedStatement = connection.prepareStatement(sql);
```

```
            if (objects != null) {
                for (int i = 0; i < objects.length; i++) {
                    preparedStatement.setObject(i + 1, objects[i]);
                }
            }
            resultSet = preparedStatement.executeQuery();
            return rsHandler.doHandler(resultSet);
        } catch (Exception e) {
            e.printStackTrace();
        } finally {
            close(connection, preparedStatement, resultSet);
        }
        return null;
    }
    //查询单个字段值通用的方法
 public static Object executeQuery(String sql, Object... objects) {
     return executeQuery(sql, new IResultSetUtil() {
         @Override
      public Object doHandler(ResultSet rs) throws SQLException {
         Object object = null;
         if (rs.next()) { object = rs.getObject(1);}
             return object;
         }
     }, objects);
    }
}
//处理 ResultSet 的接口类：IResultSetUtil.java
package com.qzmall.db;
import java.sql.ResultSet;
import java.sql.SQLException;
public interface IResultSetUtil {
    public Object doHandler(ResultSet rs) throws SQLException;
}
```

三、创建 Web 应用过滤器

基于 Java 的 Web 开发的 Servlet 有 3 类：过滤器 Filter、标准 Servlet 和监听器 Listener。

Servlet 过滤器能够对 Servlet 容器的请求和响应对象进行检查和修改。Servlet 过滤器本身并不产生请求和响应对象，只能提供过滤作用。Servlet 过滤器能够在 Servlet 被调用之前检查 Request 对象、修改 Request Header(请求头)和 Request 内容；同时，也可以在 Servlet 被调用之后检查 Response 对象、修改 Response Header 和 Response 内容。Servlet 过滤器负责过滤的 Web 组件可以是 Servlet、JSP 或者 HTML 文件。

web 应用过滤器.flv

(一)Servlet 过滤器的特点

(1) Servlet 过滤器可以检查和修改 ServletRequest 和 ServletResponse 对象。

(2) Servlet 过滤器可以被指定和特定的 URL 关联，只有当客户请求访问该 URL 时，才会触发过滤器。

(3) Servlet 过滤器可以被串联在一起，形成管道效应，协同修改请求和响应对象。

(二)使用过滤器可以完成的工作

(1) 查询请求并做出相应的活动。

(2) 阻塞"请求—响应"对，使其不能进一步传递。

(3) 修改请求的头部和数据，用户可以提供自定义的请求。

(4) 修改响应的头部和数据，用户可以通过提供定制的响应版本实现。

(5) 与外部资源进行交互。

所有的 Servlet 过滤器类都必须实现 javax.servlet.Filter 接口。这个接口包含 3 个过滤器类必须实现的方法。

(三)Servlet 过滤器创建的一般步骤

(1) 实现 javax.servlet.Filter 接口。

(2) 实现 init 方法，读取过滤器的初始化函数。

(3) 实现 doFilter 方法，完成对请求或过滤的响应。

(4) 调用 FilterChain 接口对象的 doFilter 方法，向后续的过滤器传递请求或响应。

(5) 销毁过滤器。

(四)实例讲解：设置请求和响应的编码格式

(1) 在 com.qzmall.filter 包下创建过滤器 CharsetEncodingFilter 类。代码如下：

```java
@WebFilter(filterName = "CharsetEncodingFilter",urlPatterns = "/*")
public class CharsetEncodingFilter implements Filter {
    @Override
    public void destroy() {    }
    @Override
   public void doFilter(ServletRequest request, ServletResponse response,
    FilterChain chain) throws IOException, ServletException
{ HttpServletResponse resp = (HttpServletResponse) response;
     HttpServletRequest req = (HttpServletRequest) request;
        //设置请求编码
        req.setCharacterEncoding("utf-8");
        //设置响应编码
        resp.setCharacterEncoding("utf-8");
        //清除页面缓存
        resp.setHeader("Pragma", "No-cache");
        resp.setHeader("Cache-Control", "no-cache");
        resp.setDateHeader("Expires", -10);
        chain.doFilter(request, response);   }
    @Override
 public void init(FilterConfig arg0) throws ServletException {}
        }
```

(2) 在 web.xml 中配置过滤器 CharsetEncoding。

课堂技能训练：

【实训操作内容】学习编写和配置 Servlet 过滤器来实现身份验证的方法。

【实训操作步骤】

(1) 在 com.qzmall.filter 包下创建过滤器 ValidateFilter 类。代码如下：

```java
public class ValidateFilter implements Filter {
    private ServletContext servletContext;
    @Override
    public void destroy() {  }
    @Override
    public void doFilter(ServletRequest request, ServletResponse response,
FilterChain chain) throws IOException, ServletException {
    HttpServletRequest req = (HttpServletRequest) request;
    HttpServletResponse rep = (HttpServletResponse) response;
    HttpSession session = req.getSession();
    Object user=session.getAttribute("user");
        //进行配置(用户访问的页面，都可以在 web.xml 中进行配置)
        //获取配置的登录页面地址
        String login_page = servletContext.getInitParameter("login_page");
        //获取我们需要登录才能访问的页面地址
        if (session.getAttribute("user") != null ) {
          chain.doFilter(request, response);
        } else {
          //跳转到登录页面
          System.out.println("你是需要登录才能访问的");
          rep.sendRedirect(req.getContextPath() + login_page);
        }
    }
    @Override
    public void init(FilterConfig config) throws ServletException {
        servletContext = config.getServletContext();
    }
}
```

(2) 在 web.xml 中配置过滤器 ValidateFilter。代码如下：

```xml
<context-param>
    <param-name>login_page</param-name>
    <param-value>/login.jsp</param-value>
</context-param>
<filter>
 <filter-name>validate Logn</filter-name>
    <filter-class>com.qzmall.filter.Validate Filter</filter-class>
</filter>
<filter-mapping>
    <filter-name>validate Logn</filter-name>
    <url-pattern>/cart.jsp</url-pattern>
</filter-mapping>
```

【代码说明】

在上述代码中，方法 doFilter 设置和请求的 session 对象 user 不能为空，同时设置 cart.jsp 页面需要经过上述过滤器来处理。如果 user 为空，要转向 login.jsp 页面，要求用户登录。

> **小提示**：在使用 Servlet 3.0 以上版本时，配置过滤器和 servlet 都可以采用注解的方式，比如上述的 CharsetEncodingFilter，可以在过滤方法前加上下面一段注解：
>
> @WebFilter(filterName = "CharsetEncodingFilter",urlPatterns = "/*")
>
> 这样，就没有必要再在 web.xml 中进行此过滤器配置了。同时，配置 servlet 时，也可以采用注解的方式，但必须保证是 Servlet 3.0 以上版本。但对于用户过滤器 validateLogn，需要控制的页面是有选择的，所以我们还把它统一配置在 web.xml 中。

【知识拓展】

> 过滤器在系统应用中很常见。比如在银行系统中，客户不小心把银行卡丢了，需要挂失银行卡。只要挂失银行卡，在银行系统中就不能访问这张银行卡了，这样有效地保护了客户账户的安全。这个时候就可以用过滤器来做这个应用。

课 后 训 练

一、选择题

1. 客户端发送请求后，Java 过滤器对请求的工作流程包括以下四项：
(1) 进行请求预处理；
(2) 对响应后的请求做后处理；
(3) 调用 servlet.service()对请求进行处理；
(4) 将响应结果返回给客户端。
其中为正确执行顺序的是(　　)。
　　A. (1)(2)(3)(4)　　　B. (3)(2)(1)(4)　　　C. (3)(1)(2)(4)　　　D. (1)(3)(2)(4)

2. 使用 Servlet 过滤器，需要在 web.xml 中配置(　　)元素。
　　A. <filter-mapping>　　　　　　　B. <filter-config>
　　C. <servlet-filter>　　　　　　　D. <filter>

3. servlet 过滤器参数初始化，需要在 web.xml 的 filter 元素下配置(　　)元素。
　　A. <dispatcher>　　　　　　　　B. <filter-mapping>
　　C. <url-partten>　　　　　　　　D. <init-param>

4. 相关代码如下：

```
<filter>
    <filter-name>UrlSecurityFilter</filter-name>
    <filter-class>com.qzmall.filter.UrlSecurityFilter
```

```
</filter-class>
    </filter>
    <filter-mapping>
        <filter-name>CoursewarePrivFilter</filter-name>
        <url-pattern>/courseware/*</url-pattern>
        <dispatcher>REQUEST</dispatcher>
        <dispatcher>INCLUDE</dispatcher>
</filter-mapping>
```

过滤器配置中，<dispatcher>有 request 和 include 两种方式的配置，如果此时 courseware 目录下的目标页面可以 forward 方式被调用，请问 UrlSecurityFilter 过滤器会不会被调用？（   ）

    A. 再增加一个<dispatcher>项，配置为 error，就会被调用

    B. 再增加一个<dispatcher>项，配置为 forward，就会被调用

    C. 不确定

    D. 不会

5. 通过过滤器实现编码的统一，可以通过过滤器类在 doFilter 方法下，设置(   )代码。

    A. response.setCharacterEncoding("UTF-8")

    B. request.setCharacterEncoding("UTF-8")

    C. response.setCharacterEncoding("UTF-8")

    D. request.getCharacterEncoding("UTF-8")

6. 以下是关于 Filter 生命周期的描述，横线处分别调用了哪个方法？(   )

(1) Web 应用加载后会立即创建出当前 Web 应用中的 Filter 对象；

(2) 创建后，立即调用_____方法进行初始化操作；

(3) 当关闭 Web 容器，关机，或者 reload 整个应用时，会调用_____方法来关闭 Filter。

也就是说，当 Web 容器启动时，Filter 就被加载到内存，并在销毁之前都常驻内存。

    A. destroy(),init()   B. doFilter(), destroy()

    C. init(), doFilter()   D. init(),destroy()

7. 以下哪个操作不是 Java 过滤器完成的？(   )

    A. 完成用户请求   B. 字符编码处理

    C. 权限访问控制   D. 压缩响应信息

8. 编写 Servlet 过滤器时，(   )方法用于调用过滤器链中的下一个过滤器。

    A. Servlet   B. FilterConfig

    C. Filter   D. FilterChain

9. public class SecurityFilter_____{……}

如果想要让该类成为一个过滤器，横线处应为(   )。

    A. implements HttpFilter   B. extends Filter

    C. implements Filter   D. extends HttpFilter

10. 下面哪个方法当服务器关闭时被调用，用来释放 Servlet 所占的资源？（　　）

　　A. service()　　　B. init()　　　C. doPost()　　　D. destroy()

11. 在 Java 中开发 JDBC 应用程序时使用(　　)类的 getConnection()方法可以获取 Connection 连接对象。

　　A. DBManager　　　　　　　　B. DriverManager

　　C. DBHelper　　　　　　　　　D. PreparedStatement

12. 在进行 Web 开发时经常会遇到中文乱码的问题，可以在 JSP 页面中解决乱码的方式是(　　)。

　　A. <@ page contentType="text/html;charset=中文" %>

　　B. <%@ page charset="GBK" %>

　　C. <%@ contentType="text/html;charset=GBK" %>

　　D. <%@ page contentType="text/html;charset=GBK" %>

13. 在 Web 项目的目录结构中，web.xml 文件位于(　　)中。

　　A. src 目录　　　　　　　　　B. 文档根目录

　　C. META-INF 目录　　　　　　D. WEB-INF 目录

二、实际操作题

1. 利用 navcat for mysql 数据库操作可视化工具，建立 Qzmall 数据库，按照第二节的实例，把所有表都添加进去，并设置好各个表之间的主键、外键约束，按照要求建立触发器。

2. 建立 Qzmall 数据库各表的相应实体类。

Qzmall 电子商城系统设计.pptx

第三章　习题答案.docx

第四章
jQuery 和 JSTL 标签应用

知识能力目标

1. 掌握 jQuery 和 JSTL 的基本语法。
2. 理解轮播图的实现过程。
3. 理解异步请求的含义和用途。
4. 学会 jQuery 和 JSTL 的基本配置。
5. 学会用 jQuery 实现轮播图的效果。
6. 学会使用 JSTL 实现迭代和遍历。
7. 学会异步请求的基本应用。

问题提示

现在，一般大型购物网站的首页都有轮播图，像天猫、京东、苏宁易购等。轮播图其实就是把多张图片放到网页的一个地方，按一定时间和次序进行播放。轮播图的作用是不言而喻的，它可以放网站的主推产品、热销产品、新品及其他推荐产品。单击轮播图可以链接到某一类商品的网页，让大家可以详细了解这些商品信息，从而方便消费者进行购买。

问题：轮播图的效果是如何制作出来的呢？

第一节 jQuery 简介

jQuery 是一个 JavaScript 函数库，它是一个轻量级的"写的少、做的多"的 JavaScript 库。jQuery 核心函数：$()，我们后面会一直使用这个函数。目前网络上有大量开源的 JS 框架，而 jQuery 是目前最流行的 JS 框架，它提供了大量的扩展。

jQuery 库是一个 JavaScript 文件，程序员可以使用 HTML 的 <script> 标签引用它：<script src="jquery-1.10.2.min.js"></script>。

如果程序员不希望下载并存放 jQuery，那么也可以通过 CDN(内容分发网络) 引用它，如<script src="https://libs.baidu.com/jquery/2.1.4/jquery.min.js"></script>。

一、jQuery 语法

jQuery 语法是通过选取 HTML 元素后执行一些操作。

基础语法：$(selector).action()。

(1) $表示用美元符号定义 jQuery。

(2) 选择符(selector)用于查询和查找 HTML 元素。

(3) jQuery 的 action() 执行对元素的操作。

实例：

```
$(this).hide() - 隐藏当前元素；
$("p").hide() - 隐藏所有 <p> 元素；
$("p.test").hide() - 隐藏所有 class="test" 的 <p> 元素；
$("#test").hide() - 隐藏所有 id="test" 的元素。
```

二、jQuery 选择器

jQuery 选择器允许程序员对 HTML 元素组或单个元素进行操作。

jQuery 选择器基于元素的 id、类、类型、属性、属性值等查找(或选择)HTML 元素。它主要基于已经存在的 CSS 选择器，除此之外，它还有一些自定义的选择器。

jQuery 中所有选择器都以美元符号开头：$()。

(一)元素选择器

jQuery 元素选择器基于元素名选取元素。

可在页面中选取所有 <p> 元素。

实例：用户单击按钮后，所有 <p> 元素都隐藏。代码如下：

```
$(document).ready(function(){ $("button").click(function(){ $("p").hide();
 }); });
```

(二)#id 选择器

jQuery #id 选择器通过 HTML 元素的 id 属性选取指定的元素。

页面中元素的 id 应该是唯一的，所以要在页面中选取唯一的元素需要通过#id 选择器。

通过 id 选取元素的语法为：$("#test")。

实例：当用户单击按钮后，所有 id="test" 属性的元素将被隐藏。代码如下：

```
$(document).ready(function(){ $("button").click(function(){ $("#test").
hide(); }); });
```

(三)class 选择器

jQuery 类选择器可以通过指定的 class 查找元素。

语法为：$(".test")。

实例：用户单击按钮后，所有带有 class="test" 属性的元素都隐藏。代码如下：

```
$(document).ready(function(){ $("button").click(function(){ $(".test").
hide(); }); });
```

三、jQuery 事件

页面对不同访问者的响应叫作事件。事件处理程序指的是当 HTML 中发生某些事件时所调用的方法。比如：在元素上移动鼠标、选中单选按钮、单击元素等。

在事件中经常使用术语"触发"(或"激发")。例如：当按下按键时触发 keypress 事件。常用的 jQuery 事件方法如下。

(一)$(document).ready()

$(document).ready()方法允许在文档完全加载完后执行函数。该事件方法在介绍 jQuery 语法时已经提到过。

(二)click()

click()方法是当按钮单击事件被触发时会调用一个函数。

该函数在用户单击 HTML 元素时执行。

实例：当单击事件在某个 <p> 元素上触发时，隐藏当前的 <p> 元素。代码如下：

```
$("p").click(function(){ $(this).hide(); });
```

(三)hover()

hover()方法用于模拟光标悬停事件。

当鼠标移动到元素上时，会触发指定的第一个函数(mouseenter)；当鼠标移出这个元素时，会触发指定的第二个函数(mouseleave)。

实例：

```
$("#p1").hover( function(){ alert("你进入了 p1!"); }, function(){ alert("拜拜！现在你离开了 p1!"); } );
```

(四)focus()

当元素获得焦点时，发生 focus 事件。

当通过鼠标单击选中元素或通过 Tab 键定位到元素时，该元素就会获得焦点。

focus()方法触发 focus 事件，或规定当发生 focus 事件时运行的函数。

实例：

```
$("input").focus(function(){ $(this).css("background-color","#cccccc"); });
```

(五)blur()

当元素失去焦点时，发生 blur 事件。

blur()方法触发 blur 事件，或规定当发生 blur 事件时运行的函数。

实例：

```
$("input").blur(function(){ $(this).css("background-color","#ffffff"); });
```

四、jQuery 动画

(一)用 jQuery 创建动画

jQuery animate()方法用于创建自定义动画。语法格式如下:

`$(selector).animate({params},speed,callback);`

(1) 必需的 params 参数定义形成动画的 CSS 属性。
(2) 可选的 speed 参数规定效果的时长。它可以取以下值:slow、fast 或毫秒。
(3) 可选的 callback 参数是动画完成后所执行的函数名称。

实例:演示 animate()方法的简单应用;它把 <div> 元素移动到左边,直到 left 属性等于 250 像素为止。

代码如下:

```
$("button").click(function(){
  $("div").animate({left:'250px'});
});
```

(二)用 jQuery 停止动画

jQuery stop()方法用于在动画或效果完成前对它们进行停止。stop()方法适用于所有 jQuery 效果函数,包括滑动、淡入淡出和自定义动画。

语法如下:

`$(selector).stop(stopAll, goToEnd);`

(1) 可选的 stopAll 参数规定是否应该清除动画队列。默认值是 false,即仅停止活动的动画,允许任何排入队列的动画向后执行。
(2) 可选的 goToEnd 参数规定是否立即完成当前动画。默认值是 false。

因此,默认地,stop()会清除在被选元素上指定的当前动画。

实例:演示 stop()方法,不带参数。代码如下:

```
$("#stop").click(function(){
  $("#panel").stop();
});
```

它使用 HTML5、CSS3、JavaScript 和 Ajax 通过尽可能少的代码来完成对页面的布局。

五、jQuery 遍历

(一)什么是遍历

jQuery 遍历,意为"移动",用于根据其相对于其他元素的关系来查找(或选取)HTML 元素。以某项选择开始,并沿着这个选择移动,直到抵达所期望的元素为止。

图 4-1 展示了一个家族树。通过 jQuery 遍历,能够从被选(当前的)元素开始,轻松地

在家族树中向上移动(祖先)、向下移动(子孙)、水平移动(同胞)。这种移动被称为对 DOM 进行遍历。

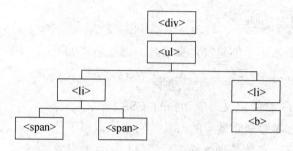

图 4-1　jQuery 遍历家族树

图示解释：

(1) <div>元素是<ul>的父元素，同时是其中所有内容的祖先。
(2) <ul>元素是<li>元素的父元素，同时是<div>的子元素
(3) 左边的<li>元素是<span>的父元素，<ul>的子元素，同时是<div>的后代。
(4) <span>元素是<li>的子元素，同时是<ul>和<div>的后代。
(5) 两个<li>元素是同胞(拥有相同的父元素)。
(6) 右边的<li>元素是<b>的父元素、<ul>的子元素，同时是<div>的后代。
(7) <b>元素是右边的<li>的子元素，同时是<ul>和<div>的后代。

提示：祖先是父、祖父、曾祖父等；后代是子、孙、曾孙等；同胞拥有相同的父。

(二)向上遍历 DOM 树

1. jQuery parent()方法

parent()方法返回被选元素的直接父元素。
该方法只会向上一级对 DOM 树进行遍历。
实例：返回每个 <span> 元素的直接父元素。代码如下：

```
$(document).ready(function(){
  $("span").parent();
});
```

2. jQuery parents()方法

parents()方法返回被选元素的所有祖先元素，它一路向上直到文档的根元素(<html>)。
实例：返回<span>元素的所有祖先。代码如下：

```
$(document).ready(function(){
  $("span").parents();
});
```

3. jQuery parentsUntil() 方法

parentsUntil()方法返回介于两个给定元素之间的所有祖先元素。

实例：返回介于\<span>与\<div>元素之间的所有祖先元素。代码如下：

```
$(document).ready(function(){
  $("span").parentsUntil("div");
});
```

(三)向下遍历 DOM 树

1. jQuery children() 方法

children() 方法返回被选元素的所有直接子元素。

该方法只会向下一级对 DOM 树进行遍历。

实例：返回每个 \<div> 元素的所有直接子元素。代码如下：

```
$(document).ready(function(){
  $("div").children();
});
```

2. jQuery find() 方法

find() 方法返回被选元素的后代元素，一路向下直到最后一个后代。

实例：返回属于 \<div> 后代的所有 \<span> 元素。代码如下：

```
$(document).ready(function(){
  $("div").find("span");
});
```

> **小提示**：什么是 jQuery Mobile
>
> jQuery Mobile 是一个为触控优化的框架，用于创建移动 Web 应用程序。
>
> jQuery 适用于所有流行的智能手机和平板电脑，jQuery Mobile 构建于 jQuery 库之上，这使其更易学习。如果程序员通晓 jQuery Mobile 的话，由它结合使用 HTML5、CSS3 和 Ajax，可通过尽可能少的代码来完成对页面的布局。

第二节 轮播图效果实现

一、轮播图页面(CSS)的实现

轮播图的界面首先要放置需要轮播的图片，图片左右两边要放置两个向左和向右的按钮，图片下面要放置几个小圆点，主要用来做图片导航。如图 4-2 所示是制作轮播图页面的静态效果。

图 4-2 轮播图页面(CSS)制作效果

页面的 HTML 及相应的 CSS 代码如下：

```html
<!DOCTYPE html>
<html>
    <head>
    <meta charset="UTF-8">
    <style type="text/css">
    *{padding:0; margin:0; list-style:none;}
    .banner{margin:100px auto;width:790px;height:300px;border:1px solid #2D8E77; position:relative;}
    .banner .img{width:50000px;position:absolute; left:0px; top:0;}
    .banner .img li{float:left;}
    .banner .num {position:absolute; width:100%;bottom:10px; left:0; text-align:center; font-size:6px; }
    .banner .num li{width:10px;height:10px; background:#888;margin:0 3px;display:inline-block; cursor:pointer; border-radius:50%;}
    .banner .num li.on{background:#f60;}
    .banner .btn{position:absolute;width:30px;heigth:50px;top:50%;margin-top:-25px;background:rgba(0,0,0,0.5); text-align:center; color:#fff; line-height:50px; font-size:40px; font-family:"宋体"; cursor:pointer;display:none;}
    .banner:hover .btn{display:block;}
    .banner .btn_l{left:0;}
    .banner .btn_r{right:0;}
    </style>
    </head>
<body>
    <div class="banner">
    <ul class="img">
        <li><a href="#"><img src="img/banner/banner1.jpg"></a></li>
        <li><a href="#"><img src="img/banner/banner2.jpg" ></a></li>
        <li><a href="#"><img src="img/banner/banner3.jpg" ></a></li>
        <li><a href="#"><img src="img/banner/banner4.jpg" ></a></li>
        <li><a href="#"><img src="img/banner/banner5.jpg" ></a></li>
    </ul>
    <ul class="num">
        <li class="on"></li>
```

```
                <li></li>
                <li></li>
                <li></li>
                <li></li>
            </ul>
            <div class="btn btn_l">&lt;</div>
            <div class="btn btn_r">&gt;</div>
        </div>
    </body>
</html>
```

二、手动单击左右按钮控制轮播的切换

下面是单击左右按钮控制轮播的切换的代码：

```
<script type="text/javascript">
    $(function(){
        var i=0
        var size=$(".banner .img li").size();
        $(".banner .num li").first().addClass("on")
        /*向左的按钮*/
        $(".banner .btn_l").click(function(){
            i++
            if(i==size){
                i=0;
            }
            $(".banner .img").stop().animate({left:-i*790},500);
            $(".banner .num li").eq(i).addClass("on").siblings().removeClass("on");
        })
        /*向右的按钮*/
        $(".banner .btn_r").click(function(){
            i--
            if(i==-1){
                i=size-1;
            }
            $(".banner .img").stop().animate({left:-i*790},500);
            $(".banner .num li").eq(i).addClass("on").siblings().removeClass("on");
        })
    })
```

运行效果如图 4-3 所示。

图 4-3　轮播效果

如果不想做无缝轮播，做到这里就可以了。

三、无缝轮播效果的实现

前面单击向左的按钮，当点到最后一张的时候，再往下单击，马上就拉到第一张，视觉效果很不好，不是想要的无缝轮播效果。那么如何才能实现无缝轮播效果呢？把第一张图复制一份，并且放到最后一张图后面，这样就变成6张图了。示例如下：

| 1 | 2 | 3 | 4 | 5 | 1 |

这样，当单击向左按钮到第五张的时候，再单击就回到第一张了，这样就完美解决了第五张图到第一张图的过渡，当再单击最后一张(即第一张图片)的时候，再把所有的图像拉到初始位置。这个效果直接用CSS来实现，并不使用animate来实现，因为用CSS来实现，并没有过渡效果，肉眼看不出来。代码如下：

```javascript
<script type="text/javascript">
    $(function(){
      var i=0
      /*将第一张图片进行复制*/
      var clone=$(".banner .img li").first().clone();
      /*粘贴到最后面*/
      $(".banner .img").append(clone)
      var size=$(".banner .img li").size();
      $(".banner .num li").first().addClass("on")
      /*向左的按钮*/
      $(".banner .btn_l").click(function(){
          i++
          if(i==size){
              $(".banner .img").css({left:0})/*回到初始位置*/
              i=1;/*回到第二张图*/
          }
          $(".banner .img").stop().animate({left:-i*790},500);
          $(".banner .num li").eq(i).addClass("on").siblings().removeClass("on");
      })
      /*向右的按钮*/
      $(".banner .btn_r").click(function(){
          i--;
          if(i==-1){
$(".banner .img").css({left:-(size-1)*790})/*回到初始位置*/
              i=size-2;
          }
          $(".banner .img").stop().animate({left:-i*790},500);
          $(".banner .num li").eq(i).addClass("on").siblings().removeClass("on");
      })
    })
</script>
```

代码设置完毕后,再把前面的 CSS 部分 banner 的样式下加上"overflow: hidden;",把溢出轮播图进行隐藏,然后浏览网页,不断单击向左按钮,可以看到所有图片都是向左的,不存在间断现象,单击向右按钮也是如此,这样就做到了无缝轮播的效果。

> **小提示:什么是 HTML5**
> HTML5 是最新的 HTML 标准,是专门为承载丰富的 Web 内容而设计的,并且无须额外插件,拥有新的语义、图形以及多媒体元素。它提供的新元素和新的 API 简化了 Web 应用程序的搭建。HTML5 是跨平台的,被设计为在不同类型的硬件(PC、平板、手机、电视机等)之上运行。HTML5 主要有以下几个新特性。
> (1) 新的语义元素,如 <header><footer><article>和<section>。
> (2) 新的表单控件,如数字、日期、时间、日历和滑块。
> (3) 强大的图像支持(借由 <canvas> 和 <svg>)。
> (4) 强大的多媒体支持(借由 <video> 和 <audio>)。
> (5) 强大的新 API,比如用本地存储取代 cookie。

课堂技能训练:

【实训操作内容】通过一个定时器来实现无缝的自动轮播和手动切换。

【实训操作要求】

(1) 无缝自动轮播。

(2) 手动切换。

(3) 圆点跟随。

提示:代码如下。

```
<script type="text/javascript">
    $(function(){
            var i=0;
            var clone=$(".banner .img li").first().clone();
            $(".banner .img").append(clone);
            var size=$(".banner .img li").size();
            //自动添加圆点
            for(var j=0;j<size-1;j++){
    $(".banner .num").append("<li></li>");
}
            $(".banner .num li").first().addClass("on");
            //单击左键
            $(".banner .btn_l").click(function(){
                movel();
            })
            //单击右键
            $(".banner .btn_r").click(function(){
              movel();
            })
            //鼠标滑入圆点
```

```
            $(".banner .num li").hover(function(){
                var index=$(this).index();
                    i=index;
        $(".banner .img").stop().animate({left:-index*790},500);
            $(this).addClass("on").siblings().removeClass("on");
                })
                //自动轮播
                var t=setInterval(movel,2000)
                //对定时器的操作
                $(".banner").hover(function(){
                    //鼠标移到图片上面时,终止定时器
                    clearInterval(t);
                },function(){
                    t=setInterval(movel,2000);
                })
                //图片向左移动函数
                function movel(){
                    i++;
                    if (i==size){
                    $(".banner .img").css({left:0});
                    i=1;
                    }
                 $(".banner .img").stop().animate({left:-i*790},500);
                    if(i==size-1){
                        $(".banner .num 
li").eq(0).addClass("on").siblings().removeClass("on");
                    }else{
                        $(".banner .num 
li").eq(i).addClass("on").siblings().removeClass("on");}
                    }
                //图片向右移动函数
            function mover(){
                    i--;
                    if (i==-1){
                    $(".banner .img").css({left:-(size-1)*790});
                    i=size-2;    }
            $(".banner .img").stop().animate({left:-i*790},500);
                $(".banner .num 
li").eq(i).addClass("on").siblings().removeClass("on");      }
                })
</script>
```

【代码解释】

(1) 首先把图片向左移动和向右移动写成两个函数,在图片向左移动的函数中,解决了前面留下的一个 bug,也就是做成无缝轮播时,因为在后面多加了图片,第一个圆点没有对应上第一张图片(也就是最后一张图片)。通过加入下面的代码把问题解决了。

```
        if(i==size-1)
{$(".banner .num li").eq(0).addClass("on").siblings().removeClass("on");
```

(2) 因为实现自动轮播,而且自动轮播图片是向左移动的,所以加了一个定时器函数 var t=setInterval(movel,2000)。

(3) 增加了鼠标滑入圆点的效果。

(4) 通过上面的例子发现,图片下面中间的圆点个数跟图片个数是一样的,为了达到更加人性化的效果,用程序自动添加圆点个数,相应的界面代码如下:

```
<ul class="num">
    <li class="on"></li>
    <li></li>
    <li></li>
    <li></li>
    <li></li>
</ul>
```

上述代码的运行结果如图 4-4 所示。

图 4-4 无缝轮播效果

【知识拓展】Tab 标签切换

提到 Tab 标签切换,大家也许并不陌生,因为在大大小小的网站中都能看到很多 Tab 切换的效果,尤其在电商类、信息类这样的网站,Tab 切换是用得非常多的。Tab 切换有一个最大的优点就是节省空间,如图 4-5 所示的这个例子就是用标签切换方式展示商品详情的一个实例。

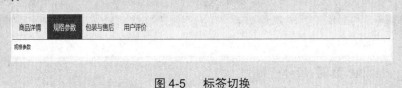

图 4-5 标签切换

第三节 Ajax 异步请求

一、Ajax 异步请求的概念

在同步请求/响应通信模型中,总是浏览器(与 Web 服务器、应用服务器或 Web 应用程序相对)发起请求(通过 Web 用户)。接着,Web 服务器、应用服务器或 Web 应用程序响应进入的请求。在处理同步请求/响应对期间,用户不能继续使用浏览器。

在异步请求/响应通信模型中,浏览器(通过 Web 用户)到 Web 服务器、应用服务器或 Web 应用程序的通信(以及反过来)是解耦的。在异步请求/响应对的处理中,Web 用户在当

异步请求.flv

前异步请求被处理时还可以继续使用浏览器。一旦异步请求处理完成，异步响应就被通信(Web 服务器、应用服务器或 Web 应用程序)反馈回客户机页面。典型情况下，在这个过程中，调用对 Web 用户没有影响，不需要等候响应。

Ajax 用来发送异步请求，请求的是服务器，但不会刷新页面。例如在注册功能中，用户在表单中输入用户名后，在用户名输入框后面会出现一个提示(用户名可用或不可用)，这说明在输入之后向服务器发出了异步请求，服务器验证这个名称是否注册过，然后返回结果，页面再通过服务器返回结果来显示用户可用或不可用的信息，但整个请求过程页面不会刷新。

jQuery 对 Ajax 提供了更方便的代码，即用$.ajax({})来发送异步请求。代码如下：

```
$(function(){
Var value=$("#xxx").val();
$.ajax({
   url : "JQueryServlet",//要请求服务器 url
   data : {method: "ajax",val:value},
//这是一个对象，它表示请求参数：method=ajax&val=xxx
    async:true,//是否为异步请求，默认为异步
    cache:false,//是否缓存结果
    type : "POST",//请求提交方式
   dataType : "json",//返回 json 格式，表示服务器返回数据是什么类型
    success : function(result) { }
//这个函数会在服务器执行成功时被调用。参数 result 就是服务器返回的值
       })
  })
```

二、异步请求在用户注册中的应用

在用户注册时，要求用户名是不能重复的，需要输入验证码，如果这两个功能完全交给后台来进行验证，对用户来说是非常不友好的。试想输入了一大堆数据，最后进行提交时，服务器提示用户不可用，那原来输入的数据又得重新输入一遍，严重影响用户的忠诚度。

(一)register.jsp 页面 html 代码

代码如下：

```
<html>
<head>
  <meta charset="utf-8" />
  <meta http-equiv="X-UA-Compatible" content="ie=edge">
  <title>会员注册</title>
<script
src="https://libs.baidu.com/jquery/2.1.4/jquery.min.js"></script>
</head>
<body>
    <div class="user-title">用户注册</div>
    <form class="form-horizontal" role="form" id="registerForm" action="addUserServlet.do"  method="post" novalidate="novalidate" onsubmit="return checkForm();" >
```

```html
    <div class="user-item">
        <label class="user-label" for="username">
          <i class="fa fa-user"></i>
        </label>
    <input class="user-content" id="username" placeholder="请输入用户名"
      name="username"  onblur="checkUsername();"
onfocus="checkHideUsername();" />
      <span id="span1" >用户名</span>
            </div>
      <button type="submit" class=" btn btn-submit">注册</button>
        </form>
      <div class="link-item">
        <a href="login.jsp" class="link" >已有账号,去登录</a>
     </div>
    </div>
</body>
</html>
```

【代码说明】

onblur="checkUsername();" 和 onfocus="checkHideUsername();"函数的作用是当用户失去用户名输入框焦点时,会出现提示信息,当用户获得焦点时提示信息隐藏。

(二)使用 jQuery 进行异步请求

register.js 代码如下:

```javascript
    //异步请求验证用户名是否存在
function checkUsername(){
    var username =$("#username"); // 获得用户名文本框的值:
    var patten =/^[a-zA-Z]\w{3,19}$/;
    if(!patten.test(username.val())) {
        alert("只能为英文或数字,且不能以数字打头,长度为4-20个字符!");
        return false;
    }
    $.ajax({
        url : "userCheckServlet.do",
        type : "POST",
        data : {
            "username" : username.val(),
            "time" : new Date().getTime()
            //因为ajax有延迟,所以还要传一个时间参数
        },
        dataType : "json",
        success : function(result) {
            if (result == 1) {
                $("#span1").css('display','inline-block');
                $("#span1").html("用户名已存在!");
                return false;
            }
            else {
                $("#span1").css('display','inline-block')
                $("#span1").html("用户名可以使用")
            }
        },
```

```
        error : function(er) {
            console.log(er);
        }
    });
}
function checkHideUsername(){
    $("#span1").css('display','none')
}
```

(三)数据访问层设计

1. 在 com.qzmall.dao.IUserDao 接口类中建立方法

代码如下:

```
public interface IUserDao {
//查询用户信息是否重复
public boolean getUserByUsername(String username);
}
```

2. 在 com.qzmall.dao.UserDaoImpl 类下实现其方法

代码如下:

```
public class UserDaoImpl implements IUserDao {
@Override
public boolean getUserByUsername(String username) {
 String strSql = "select * from user where username=?";
   return (boolean)DBUtil.executeQuery(strSql, new IResultSetUtil()
{ public Object doHandler(ResultSet rs) throws SQLException {
      if (rs.next()) { return true;}
            return false;
        }
    },username);  }
 }
```

(四)业务逻辑层设计

1. 在 com.qzmall.service.IUserService 中建立方法

代码如下:

```
public interface IUserService {
//查询用户信息是否重复
public boolean getUserByUsername(String username);}
```

2. 在 com.qzmall.service.UserServiceImpl 中实现其方法

代码如下:

```
public class UserServiceImpl implements IUserService {
@Override
   public boolean getUserByUsername(String username) {
      IUserDao userDao=new UserDaoImpl();
```

```
        return userDao.getUserByUsername(username);   }
}
```

(五)控制层设计

在 com.qzmall.servlet.user 包下建立 UserCheckServlet 类，代码如下：

```
@WebServlet(name = "UserCheckServlet")
public class UserCheckServlet extends HttpServlet {
    private static final long serialVersionUID = 1L;
    //用来表明类的不同版本间的兼容性
    @Override
    protected void doPost(HttpServletRequest req, HttpServletResponse resp)
        throws ServletException, IOException {
        String username = req.getParameter("username");
        String result="";
        IUserService userService = new UserServiceImpl();
        if (userService.getUserByUsername(username)){
            result="1"; }
        else{
            result="2"; }
        resp.getWriter().print(result);
        PrintWriter out = resp.getWriter();
    }
}
```

【代码说明】

该 UserCheckServlet 首先接受异步请求传过来的用户名，然后调用 userService.getUserByUsername (username)方法验证用户名是否存在，如果存在，result="1"，否则 result="2"，然后把 result 值返回客户端。

运行效果如图 4-6 所示。

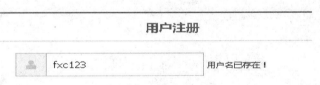

图 4-6 提示用户名已经存在

课堂技能训练：

【实训操作内容】完成异步请求验证用户名是否存在。

【实训操作步骤】

(1) 持久层设计。

(2) 业务逻辑层设计。

(3) 控制层设计。

(4) 视图层设计(jQuery 代码实现，重点)。

小提示：利用 IntelliJ IDEA 自动优化导入包

在写 Java 代码的时候，经常要引用自己写的包或 Java 自带的包，如果这个全部用手工来写的话，会大大降低开发效率，IntelliJ IDEA 带有自动导入包功能。打开 IDEA，选择 File-Settings→Editor→General→Auto Import 选项，进入如图 4-7 所示的界面。

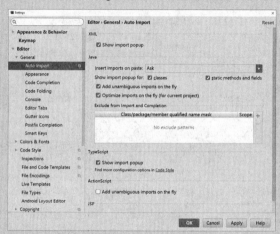

图 4-7　IDEA 自动导入包

选中 Optimize imports on the fly 和 Add unambiguous imports on the fly 复选框。

(1) Optimize imports on the fly：自动去掉一些没有用到的包；

(2) Add unambiguous imports on the fly：自动帮我们优化导入的包。

最后单击 OK 按钮即可。

【知识拓展】

异步请求在网站中的应用非常多，比如用户注册时需要输入验证码，那么验证输入正确或错误的验证最好先通过异步请求的方式来进行。还有在购物车里的商品数量更新时，小计金额、总计金额都会变动，最好也采用异步请求的方式来实现。

第四节　JSTL 常用标签

jstl 标签.flv

一、EL 表达式介绍

Expression Language(表达式语言)目的是替代 JSP 页面中的复杂代码。

EL 表达式语法如下：

${变量名}

二、JSTL 标签应用

(一)为什么要使用 JSTL 标签

简化了 JSP 的开发，减少了 JSP 中的 Scriptlet 代码数量。

EL 表达式虽然可以解决"不用书写 Java 代码"的问题,但是对于复杂的数据(如数组、集合等)取值会很麻烦。

使用 JSTL 标签配合 EL 表达式能够很好地解决复杂类型数据解析的问题,简化代码的书写。

(二)JSTL 标签的具体应用

1. JSTL 开发准备

在 JSP 页面中添加 taglib 指令:

```
<%@ taglib uri="http://java.sun.com/jsp/jstl/core" prefix="c"%>
```

2. JSTL 常用标签介绍

(1) 通用标签:set、out、remove。

① set 标签:将值保存到指定范围里。

例如:将 value 值存储到范围为 scope 的变量 variable 中:

```
<c:set var="username" value="张三" scope=" scope "/>
```

② out 标签:将结果输出显示。

例如:

```
<c:out value="value" />
```

③ remove 标签:删除指定域内数据。

例如:

```
<c:remove var="username" scope="session"/>
```

(2) 条件标签:if、choose。

实例:

```
<%@ page language="java" contentType="text/html; charset=utf-8"
    pageEncoding="utf-8"%>
<%@ taglib uri="http://java.sun.com/jsp/jstl/core"  prefix="c" %>
<html>
<head>
<meta http-equiv="Content-Type" content="text/html; charset=utf-8">
<title>jstl 中的 if 标签和 choose 标签</title>
</head>
<body>
    <c:set var="age" value="12" scope="request"></c:set>
    <!-- if 标签:
        test:根据判断的条件,如果条件为 true,则执行标签体中的内容
    -->
    <%-- <c:if test="${age==12 }">
            您的年龄为 12 岁
    </c:if>
    hello world --%>
```

```
        <hr>
        <!-- choose 标签 -->
        <c:choose>
            <c:when test="${age==12 }">
            您的年龄为 12 岁
            </c:when>
            <c:otherwise>
            您的年龄不为 12 岁
            </c:otherwise>
        </c:choose>
</body>
</html>
```

(3) 迭代标签：forEach。

forEach 标签的语法定义如下：

```
<c:forEach var="name" items="expression" varStatus="name"
 begin="expression" end="expression" step="expression">
 body content
</c:forEach>
```

- var：迭代参数的名称。在迭代体中可以使用的变量的名称，用来表示每一个迭代变量。类型为 String。
- items：要进行迭代的集合。对于它所支持的类型将在下面进行讲解。
- varStatus：迭代变量的名称，用来表示迭代的状态，可以访问到迭代自身的信息。
- begin：如果指定了 items，那么迭代就从 items[begin]开始进行迭代；如果没有指定 items，那么就从 begin 开始迭代。它的类型为整数。
- end：如果指定了 items，那么就在 items[end]结束迭代；如果没有指定 items，那么就在 end 结束迭代。它的类型也为整数。
- step：迭代的步长。

(三)利用 forEach 标签对 list 集合进行迭代

(1) 在 com.qzmall.servlet.jstl 建立 JstlServlet，代码如下：

```
@WebServlet(name = "JstlServlet",urlPatterns ="/jstl.do" )
public class JstlServlet extends HttpServlet {
    private static final long serialVersionUID = 1L;
    //用来表明类的不同版本间的兼容性
    protected void doGet(HttpServletRequest req, HttpServletResponse resp)
throws ServletException, IOException {
      Map<String,Object> dataMap1=new HashMap<String,Object>();
      dataMap1.put("shopName","联想笔记本");
      dataMap1.put("address","北京");
      dataMap1.put("price",4999.99);
      Map<String,Object> dataMap2=new HashMap<String,Object>();
      dataMap2.put("shopName","神舟笔记");
      dataMap2.put("address","南京");
      dataMap2.put("price",3999.99);
```

```
        Map<String,Object> dataMap3=new HashMap<String,Object>();
        dataMap3.put("shopName","苹果笔记本");
        dataMap3.put("address","上海");
        dataMap3.put("price",7999.99);
        List<Map<String,Object>> mapList=new ArrayList<Map<String,Object>>();
        mapList.add(dataMap1);
        mapList.add(dataMap2);
        mapList.add(dataMap3);
        //将存放多条数据的 mapList 集合保存到 req 域中
        req.setAttribute("mapList",mapList);
        //将 mapList 集合发送到 jstllist.jsp
req.getRequestDispatcher("jstllist.jsp").forward(req,resp);
    }   }
```

(2) 在 Web 目录下建立 jstllist.jsp，代码如下：

```
<%@ page language="java" contentType="text/html; charset=utf-8"
    pageEncoding="utf-8"%>
<%@ taglib uri="http://java.sun.com/jsp/jstl/core" prefix="c"%>
<html>
<head>
<meta http-equiv="Content-Type" content="text/html; charset=utf-8">
<title>通过 jstl 和 EL 表达式迭代 List 集合</title>
</head>
<body>
    <center>
        <table border="1px" cellspacing="0px">
            <tr>
                <td>商品名称</td>
                <td>产地</td>
                <td>价格</td>
            </tr>
            <c:forEach items="${mapList }" var="Map">
                <tr>
                    <td>${Map.shopName}</td>
                    <td>${Map.address}</td>
                    <td>${Map.price}</td>
                </tr>
            </c:forEach>
        </table>
    </center>
</body>
</html>
```

(3) 打开浏览器，在地址栏中输入 http://localhost:8080/qzmall/jstl.do，运行效果如图 4-8 所示。

图 4-8 利用 forEach 标签迭代数据

课后训练

一、选择题

1. 以下关于 EL 表达式，说法错误的是( )。

   A. EL 表达式通常用在 JSP 页面中，目的是使 JSP 代码更加简洁

   B. EL 表达式的全称是 Expression Language

   C. ${8+5}的返回值是 13

   D. EL 表达式可以用到 HTML 页面中

2. 将存储在 request 域中的变量 username 值取出，下列写法正确的是( )。

   A. <c:set value= "${username}"></c:set>

   B. <c:out value= "${username}"></c:out>

   C. <c:set var= "username" value= "Smith" Scope= "request">

   D. <c:out value= "username"></c:out>

3. 已知 JSP 页面的代码如下，此时页面的输出结果是( )。

```
<c:set var="score" value="85"></c:set>
<c:choose>
    <c:when test="${score>=90}">你的成绩为优秀！</c:when>
    <c:when test="${score>=70&&score<90}">你的成绩为良好！</c:when>
    <c:when test="${score>=60&&score<70}">你的成绩为及格！</c:when>
    <c:otherwise>
        对不起，你没有通过考试
    </c:otherwise>
</c:choose>
```

   A. 你的成绩为及格！   B. 你的成绩为良好！

   C. 你的成绩为优秀！   D. 对不起，你没有通过考试

4.

```
<c:forEach var="name" items="collection" begin="1" end="n" step="step">
  主体内容
</c:forEach>
```

<c:forEach>迭代标签的格式，其中 item 的类型可以是( )。

   A. 其余选项都正确   B. Iterator   C. Arrays

   D. Set   E. Map

5.

```
<%
pageContext.setAttribute("colors", new String[]{"red","blue","green","pink","dark"});
%>
```

```
<c: var="color" items="${colors}">
    \${color}=${color}<br/>
</c:forEach>
```

为了能够显示所有颜色数据,需要在颜色处用到的标签是( )。

  A. forEach B. when C. choose D. out

6. 与 Java 中增强 for 循环对应的 JSTL 标签是( )。

  A. <c:forEach> B. <c:for> C. <c:when> D. <c:if>

7.
```
<c:set var="sum" value="0" scope="request"></c:set>
<c:forEach begin="1" end="100" step="1" varStatus="st">
    <c:set var="sum" value="    "></c:set>
</c:forEach>
<c:out value="${sum}" />
```

上述代码使用 JSTL 标准标签库中的标签输出 1 到 100 中所有的数字之和。应该填写的代码是( )。

  A. $[sum+st.count] B. {sum+st.count}

  C. ${sum+st.count} D. sum+st.count

8. 在使用 JSTL 进行 JSP 页面开发过程中,需要在 JSP 页面中添加 taglib 指令,以下选项中哪一项是错误的?( )

  A. <%@ taglib url="http://java.sun.som/jsp/jstl/core" prefix="c"%>

  B. <%@ taglib url="http://java.sun.som/jsp/core" prefix="e"%>

  C. <%@ taglib url="http://java.sun.som/jsp/jstl/core" prefix="e"%>

  D. <%@ prefix="d" taglib url="http://java.sun.som/jsp/jstl/core" %>

9. 在 JSP 页面上使用 JSTL 标签时,应使用 taglib 指令导入标签库描述符文件,并设置 taglib 指令的( )属性指定标签的前缀。

  A. prefix B. name C. tag D. url

10. Ajax 中的属性 type 代表的含义是( )。

  A. 请求时数据的传递方式 B. 交互成功后要执行的方法

  C. 接收后台的数据类型 D. 当前页面的类型

11. 在对象 XMLHttpRequest 的属性 status 值为( )时表示请求响应一切正常。

  A. 500 B. 200 C. 404 D. 403

12. 在使用 Ajax 实现页面处理时,Servlet 给 Ajax 返回数据的数据格式为 json,需要在( )属性中表现出来。

  A. datatype B. data C. json D. url

13. 对象 XMLHttpRequest 的属性 readyStatus 值为( )时表示请求已完成且响应已就绪。

  A. 4 B. 1 C. 2 D. 3

二、实际操作题

1. 编写 jQuery 代码,实现网页页面导航功能。
2. 编写 jQeury 代码,实现购物车前台的选择商品、小计和总计功能。

第四章 jQuery 和 JSTL 标签应用.pptx

第四章 习题答案.docx

第五章
商品显示模块设计

知识能力目标

1. 了解商品信息主要包括哪些内容。
2. 了解商品显示主要包括哪些方面。
3. 学会最新商品显示编码设计。
4. 学会分类分页商品显示编码设计。
5. 学会商品详情显示编码设计。
6. 学会商品模糊查询编码设计。

问题提示

在现实生活中，当逛乐购、家乐福这样的大型超市时，通常会通过超市的指示牌清晰地知道想买的商品大概在什么位置。在节日或者活动时都不需要走多远，在超市进门处就会发现售卖节日相关商品的货架。

问题：在电子商城中，如何让用户在繁杂的类目中找到想要的商品？

第一节 首页子模块设计

一、首页子模块功能分析

首页子模块主要功能有：商品广告轮播图显示；商品类别显示；页面导航效果。

商品广告轮播图显示主要是把 ad 表里的最新前 5 件商品图片显示，根据每个商品指向的类别再链接到分类页面。轮播图效果前面已讲过。

商品类别主要把商品大类及子类表里的数据按照层级关系显示出来，单击相应类别链接到各分类页面。

页面导航主要是在页面滚动鼠标时能和右边导航条焦点相对应，单击导航条的商品大类能够自动切换到该商品大类下的最新 10 件商品，导航效果由 jQuery 代码编写。

首页页面设计.flv

首页子模块设计.flv

二、基础知识

(一) 调用 Servlet 的主要方法

(1) forward 动作调用：<jsp:forward page="index.do"/>。

(2) 表单调用：<form action="addUserServlet.do" method="post">。

(3) 超级链接调用：<a href="index.do">首页</a>。

(二) Servlet 请求转发、重定向

(1) forward：是指转发，将当前 request 和 response 对象保存，交给指定的 url 处理。

并没有表示页面的跳转,所以地址栏的信息不会发生改变。代码如下:

```
req.getRequestDispatcher("/success.html").forward(req, resp);
```

(2) SendR:是指重定向。包含两次浏览器请求,浏览器根据 url 请求一个新的页面,所有的业务处理都转到下一个页面,地址栏的信息会发生改变。代码如下:

```
resp.SendRedirect(req.getContextPath()+"/fail.html");
```

三、首页子模块功能实现

(一)首页子模块持久层设计

1. 接口类设计

在 com.qzmall.dao 包下新建 IShopDao 接口类:

```java
public interface IShopDao {
    //查询所有分类
    public ArrayList<Category> getCategoryAll();
    //查询所有大类最新商品
    public ArrayList<Category> getCategoryShopAll();
    //查询首页广告
    public List<Ad> adlist();
    //查询所有大类
    public List<Category> getCategory();
}
```

2. 接口实现类设计

在 com.qzmall.dao.impl 包下新建 ShopDaoImpl 类:

```java
public class ShopDaoImpl implements IShopDao {
/*
```

(1) 该方法是查询所有分类,返回结果为集合类型。代码如下:

```java
*/
@Override
public ArrayList<Category> getCategoryAll() {
    Connection conn = DBUtil.getConnection();
    PreparedStatement pstmt = null;
    PreparedStatement psub = null;
    ResultSet rs = null;
    ResultSet rsub = null;
    ArrayList<Category> categoryArrayList = new ArrayList<Category>();
    ArrayList<SubCate> subCateArrayList = new ArrayList<SubCate>();
    try {
        String sql = "select * from category ";
        pstmt = conn.prepareStatement(sql);
        rs = pstmt.executeQuery();
        while (rs.next()) {
```

```
            Category category = new Category();
            category.setCid(rs.getInt("cid"));
            category.setCname(rs.getString("cname"));
            String strsql = "select * from subcate where cid=" +
rs.getInt("cid");
            //向每一个大类添加它的子类
            psub = conn.prepareStatement(strsql);
            rsub = psub.executeQuery();
            while (rsub.next()) {
                SubCate subCate = new SubCate();
                subCate.setSid(rsub.getInt("sid"));
                subCate.setSname(rsub.getString("sname"));
                subCate.setCategory(category);
                subCateArrayList.add(subCate);
            }
            category.setSubCates(subCateArrayList);
            categoryArrayList.add(category);
        }
        return categoryArrayList;
    } catch (SQLException e) {
        // TODO Auto-generated catch block
        e.printStackTrace();
        return null;
    } finally {
        DBUtil.close(conn, pstmt, rs);
        DBUtil.close(conn, psub, rsub);
    }
}
/*
```

【代码说明】

该方法用来实现查询每个大类下所有子类，首先遍历所有大类，然后往每个大类下添加属于它的子类，返回结果为 ArrayList<Category> 类型。

(2) 查询所有大类下的最新商品。代码如下：

```
*/
@Override
public ArrayList<Category> getCategoryShopAll() {
    Connection conn = null;
    conn = DBUtil.getConnection();
    PreparedStatement pstmt = null;
    PreparedStatement pshop = null;
    ResultSet rs = null;
    ResultSet rshop = null;
    ArrayList<Category> categories = new ArrayList<Category>(0);
    ArrayList<Shop> shopArrayList = new ArrayList<Shop>(0);
    try {
        String sql = "select * from category ";
        pstmt = conn.prepareStatement(sql);
        rs = pstmt.executeQuery();
```

```
            while (rs.next()) {
                Category category = new Category();
                category.setCid(rs.getInt("cid"));
                category.setCname(rs.getString("cname"));
                String strsql = "select * from shop where cid=" +
rs.getInt("cid") + " order by shopdate desc limit 0,10 ";
                pshop = conn.prepareStatement(strsql);
                rshop = pshop.executeQuery();
                while (rshop.next()) {
                    Shop shop = new Shop();
                    shop.setPid(rshop.getInt("pid"));
shop.setShopname(rshop.getString("shopname"));
                    shop.setImage1(rshop.getString("image1"));
                    shop.setCategory(category);
                    shopArrayList.add(shop);
                }
                category.setShopArrayList(shopArrayList);
                categories.add(category);
            }
            return categories;
        } catch (SQLException e) {
            // TODO Auto-generated catch block
            e.printStackTrace();
            return null;
        } finally {
            DBUtil.close(conn, pshop, rshop);
            DBUtil.close(conn, pstmt, rs);
        }
    }
/*
```

【代码说明】

该方法用来查询每个大类下的最新商品,首先遍历所有大类,然后往每个大类添加 10 件最新商品,返回结果为 ArrayList<Category>类型。

(3) 首页轮播图。代码如下:

```
*/
    @Override
    public List<Ad> adlist() {
    List<Ad> adlist = new ArrayList<Ad>();
    String strSql = "select subcate.sname as sname,ad.* from ad,subcate
where ad.sid=subcate.sid order by ad.atime desc limit 0,5";
        return (List<Ad>) DBUtil.executeQuery(strSql, new IResultSetUtil() {
        @Override
         public Object doHandler(ResultSet rs) throws SQLException {
            while (rs.next()) {
                Ad ad = new Ad();
                ad.setAimage(rs.getString("aimage"));
                ad.setSname(rs.getString("sname"));
                ad.setSid(rs.getInt("sid"));
```

```
            ad.setAtime(rs.getString("atime"));
            adlist.add(ad);
         }
       return adlist; }
    });
}
/*
```

【代码说明】

由于进行轮播的商品图片只需要最新的 5 张，所以只需要查询 ad 表中的按时间排序的前 5 条记录即可。由于轮播图片是为某一子类商品做广告的，所以单击轮播图片返回的结果应是显示所有子类的商品首页。因此上面的 SQL 语句的功能还有查询轮播图片是属于哪个商品子类的功能。

(4) 查询所有大类。代码如下：

```
*/
@Override
public List<Category> getCategory() {
List<Category> categoryArrayList = new ArrayList<Category>();
String sql = "select * from category ";
return ( List<Category>)DBUtil.executeQuery(sql, new IResultSetUtil() {
      public Object doHandler(ResultSet rs) throws SQLException {
        while (rs.next()) {
            Category category = new Category();
            category.setCid(rs.getInt("cid"));
            category.setCname(rs.getString("cname"));
            categoryArrayList.add(category);}
            return categoryArrayList; }
      });
   }
}
```

小提示：用 IDEA 自动添加接口实现类方法

用 IDEA 添加接口实现类，非常方便，当进入添加实现类界面，用鼠标选中类名称，然后按 Alt+Enter 组合键，选择第一项，如图 5-1 所示，就可以有选择地或全部把接口类中的方法加载进去。加载其他类方法也是如此。

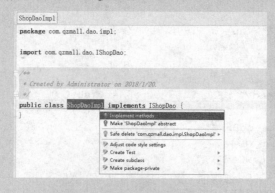

图 5-1　添加接口实现类方法

(二)首页子模块业务逻辑层设计

1. 接口类设计

在 com.qzmall.service 包下建立接口类 IShopService 类。代码如下:

```java
public interface IShopService {
  //查询所有分类
  public ArrayList<Category> getCategoryAll();
   //查询所有大类最新商品
  public ArrayList<Category> getCategoryShopAll();
   //查询首页广告
  public List<Ad> adlist();
  // 查询所有大类
  public List<Category> getCategory();  }
```

2. 接口实现类设计

在 com.qzmall.service.impl 包下建立接口实现类 ShopServiceImpl,用以实现接口类的方法。代码如下:

```java
public class ShopServiceImpl implements IShopService {
    @Override   //查询所有分类
    public ArrayList<Category> getCategoryAll() {
        IShopDao shopDao=new ShopDaoImpl();
        return shopDao.getCategoryAll();  }
    @Override  //查询所有大类最新商品
    public ArrayList<Category> getCategoryShopAll() {
        IShopDao shopDao=new ShopDaoImpl();
        return shopDao.getCategoryShopAll();}
    @Override   //显示首页广告
    public List<Ad> adlist() {
        IShopDao shopDao=new ShopDaoImpl();
        return shopDao.adlist();}
    @Override   //显示所有大类
    public List<Category> getCategory() {
        IShopDao shopDao=new ShopDaoImpl();
        return shopDao.getCategory();  }
}
```

(三)首页子模块控制层设计

商品显示模块控制层是连接显示层和持久层的桥梁,可以分派用户的请求并选择恰当的视图以用于显示,同时可以解释用户的输入并映射为模型层可执行的操作。在 com.qzmall.servlet.shop 类下建立 servlet:CategoryAllServlet,具体代码如下:

```java
@WebServlet(name = "CategoryAllServlet",urlPatterns = "/index.do")
public class CategoryAllServlet extends HttpServlet {
    protected void doGet(HttpServletRequest req, HttpServletResponse resp) throws ServletException, IOException {
        IShopService categoryShopService=new ShopServiceImpl();
```

```
        List<Category> categoryList= categoryShopService.getCategory();
        req.setAttribute("categoryList",categoryList);
        List<Ad> adList=categoryShopService.adlist();
        req.setAttribute("adList",adList);
        ArrayList<Category>
categoryShopArrayList=categoryShopService.getCategoryShopAll();
req.setAttribute("categoryShopArrayList",categoryShopArrayList);
        ArrayList<Category>
categoryArrayList=categoryShopService.getCategoryAll();
        req.setAttribute("categoryArrayList",categoryArrayList);
        req.getRequestDispatcher("main.jsp").forward(req,resp);
    }
}
```

【代码说明】

CategoryAllServlet 通过调用商品业务逻辑对象 categoryShopService 实现以下四个方面的功能。

(1) 通过商品业务对象 categoryShopService 的 adlist 方法生成首页轮播图集合 adList，并回传到主页(main.jsp)。

(2) 通过商品业务对象 categoryShopService 的 getCategoryAll()方法生成商品类别集合 categoryList，并回传到主页(main.jsp)。

(3) 通过商品业务对象 categoryShopService 的 getCategoryShopAll()方法生成大类商品集合 categoryShopArrayList，并回传到主页(main.jsp)。

(4) 通过(3)后，再用于进行页面导航。

(四)首页子模块视图层设计

1. index.jsp 页面设计

代码如下：

```
<%@ page contentType="text/html;charset=UTF-8" language="java" %>
<jsp:forward page="index.do"/>
```

2. 头部(head.jsp)页面设计

代码如下：

```
<%@ page contentType="text/html;charset=UTF-8" language="java" %>
<!DOCTYPE html>
<html>
<head>
    <link type="text/css" rel="stylesheet" href="css/font-awesome.css">
    <link type="text/css" rel="stylesheet" href="css/reset.css">
    <link type="text/css" rel="stylesheet" href="css/head.css">
</head>
<body>
<div class="nav">
    <div class="w">
```

```html
        <div class="user-info">
        <c:choose>
        <c:when test="${empty sessionScope.user }">
        <span ><a  class="link js-login"href="login.jsp">登录</a> </span>
        <span ><a class="link js-register" href="register.jsp">注册</a></span>
        </c:when>
        <c:otherwise>
        <span class="login">
        <span class="link-text">
        <span class="link-text">欢迎你,
        <span class="username">${sessionScope.user.truename}</span>
         </span>
        <span><a href="quit.do"  class="link">退出</a> </span>
         </span>
        </c:otherwise>
        </c:choose>
          </div>
         <ul class="nav-list">
           <li class="nav-item">
           <a class="link" href="ordersInfo.do?currentPage=1">我的订单</a>
            </li>
            <li class="nav-item">
             <a class="link" href="userInfo.do">个人中心</a>
             </li>
             </ul>
      </div></div>
<div class="header">
 <div class="w">
  <a href="index.do" class="link logo">QZMALL</a>
   <div class="search-con">
   <form action="queryShop.do?" method="get" onsubmit="return check();">
    <input class="search-input" id="shopname" name="shopname" placeholder="请输入商品名称"/>
    <input class="currentPage" id="currentPage" name="currentPage" value="1" type="hidden"/>
      <input type="submit" class="btn search-btn" id="search-btn" value="搜索"></input>
       </form>
        </div>
      </div>
</div>
</body>
<script type="text/javascript">
    function check(){
        var checkyzm = $("#shopname");
       if (checkyzm.val() == "") {
          alert("商品名称不能为空! ");
          checkyzm.focus();
          return false; }
    }
```

3. main.jsp 主界面设计

代码如下：

```jsp
<%@ taglib prefix="c" uri="http://java.sun.com/jsp/jstl/core" %>
<%@ page contentType="text/html;charset=UTF-8" language="java" %>
<!DOCTYPE html>
<html>
<head>
  <meta charset="UTF-8">
  <meta http-equiv="X-UA-Compatible" content="ie=edge">
  <title>秦职电商平台</title>
  <link type="text/css" rel="stylesheet" href="css/font-awesome.css">
  <link type="text/css" rel="stylesheet" href="css/reset.css">
  <link type="text/css" rel="stylesheet" href="css/index.css">
  <script src="https://libs.baidu.com/jquery/2.1.4/jquery.min.js"></script>
  <script src="js/focus.js"></script>
  <script src="js/pageNav.js"></script>
</head>
<body>
<jsp:include page="head.jsp"/>
<div class="w">
  <!--左侧导航-->
  <ul class="keywords-list">
    <c:forEach items="${categoryArrayList}" var="category">
    <li class="keywords-item">
      <dl class="out">
        <dt>
        <a href="cateList.do?cid=${category.cid}&cname=${category.cname}&currentPage =1" target="_blank">${category.cname}</a>
        </dt>
        <dd>
          <c:forEach items="${category.subCates}" var="sname">
            <c:if test="${category.cid==sname.category.cid}">
            <a href="subList.do?sid=${sname.sid}&cname=${sname.sname}&currentPage=1" target="_blank">${sname.sname}</a>
            </c:if>
          </c:forEach>
        </dd>
      </dl>
    </li>
    </c:forEach>
  </ul>
  <!--轮播图-->
  <div class="banner">
    <ul class="img">
    <c:forEach items="${adList}" var="ad">
```

```
            <li><a href="subList.do?sid=${ad.sid}&cname=${ad.sname}&currentPage
=1"><img src="<c:url value='img/banner/${ad.aimage}'/>" height="370"
width="830"/></a></li>
        </c:forEach>
        </ul>
        <ul class="num">
        </ul>
        <div class="btn btn_l">&lt;</div>
        <div class="btn btn_r">&gt;</div>
    </div>
  </div>
  <!--导航及轮播图结束-->
<!--右侧分类导航开始-->
  <div id="menu">
    <ul>
    <c:set var="i" value="1" />
<c:forEach items="${categoryList}" var="ca">
      <li><a href="#item${i}">F${i} ${ca.cname}</a></li>
      <c:set value="${i + 1}" var="i" />
</c:forEach>
    </ul>
  </div>
<!--右侧分类导航结束-->
  <!--楼层开始-->
  <div class="w">
    <c:set var="i" value="1" />
<c:forEach items="${categoryShopArrayList}" var="cate">
    <div class="floor-wrap" id="item${i}">
      <h1 class="floor-title">F${i}  ${cate.cname}</h1>
      <ul class="floor-list">
      <c:forEach items="${cate.shopArrayList}" var="shop">
      <c:if test="${cate.cid==shop.category.cid}">
        <li class="floor-item">
          <a href="" >
          <span class="floor-text">${shop.shopname}</span>
          <a href="detail.do?pid=${shop.pid}"><img src="<c:url
value='img/pic/${shop.image1 }'/>" class="floor-img"></a>
          </a>
        </li>
      </c:if>
   </c:forEach>
      </ul>
    </div>
    <c:set value="${i + 1}" var="i" />
</c:forEach>
  </div>
  <!--楼层结束-->
  <jsp:include page="foot.jsp"/>
  </body>
</html>
```

【代码说明】

(1) 为了维护代码的统一性，专门做了统一的页头(head.jsp)和页脚(foot.jsp)页面。

(2) 左侧分类导航遍历，利用 JSTL 标签对传来的 categoryArrayList 集合进行遍历。先遍历出大类，然后查找该大类下的子类进行遍历。

(3) 大类商品遍历，利用 JSTL 标签对传来的 categoryShopArrayList 集合进行遍历，先遍历出大类名称，再查询大类下的商品。由于需要分楼层遍历出每个大类最新 10 件商品，在遍历时不但要遍历出每个大类下的商品，还需要各个楼层标签，比如：F1 家用电器，所以在遍历前通过 JSTL 标签设置了变量 i，并赋初值等于 1，每往下循环一个大类，变量 i 加 1。

(4) 轮播图遍历。利用 JSTL 标签对传来的 adList 集合进行遍历。

(5) 右侧导航条遍历。利用 JSTL 标签对传来的 categoryList 集合进行遍历。

(6) 轮播图和网页导航的 jQuery 代码请参照第四章 jQuery 应用部分。

运行效果如图 5-2 所示。

图 5-2　首页运行效果

📝 **课堂技能训练:**

【实训操作内容】完成首页大类商品显示功能设计。

【实训操作步骤】

(1) 持久层设计(重点)。
(2) 业务逻辑层设计。
(3) 控制层设计。
(4) 视图层设计。

第二节 商品分类显示子模块设计

一、商品分类显示子模块功能分析

(1) 根据查询条件查询所有符合条件的商品信息。
(2) 在首页单击商品大类或子类,可以显示某大类下或某子类的商品列表信息。
(3) 在页面上单击某一图片,能够链接到商品详情页。

商品分类显示.flv

二、分页查询要求的基本数据

(1) 当前页数据:shopList 通过调用商品业务逻辑类来求得。
(2) 当前页页码:currentPage,默认为 1,如果页面传递了值,就使用页面传递的。
(3) 当前页记录:使用 sql 语句 "limit ?,?",第一个问号代表起始记录,第二个问号代表当前页显示多少条记录。
(4) 总页数:totalPage,使用总记录数(totalSize)和每页记录数(Tally.SHOP_PAGE_SIZE)来求得。
(5) 为了保证查询条件不丢,还需要记录第一次查询的条件:请求的 servlet 路径及参数:

```
<a href="${url}&currentPage=$(currentPage-1)>上一页</a>"
```

上面的 url 通过 req.getRequestURL()+"?"+req.getQueryString()参数求得。

三、商品分类显示子模块功能实现

(一)商品分类显示子模块持久层设计

1. 接口类设计

在 com.qzmall.dao.IShopDao 类下添加下列方法:

```
public interface IShopDao {
    //分页查询大类商品
```

```
public List<Shop> getShopCategoryPage(int cid, int currentPage, int pageSize);
public int getTotalByCid(int cid);
 //按商品名称分页模糊查询
public List<Shop> getShopByShopname(String shopname,int currentPage,int pageSize);
public int getTotalByShopname(String shopname);
}
```

2. 接口实现类设计

在 com.qzmall.dao.impl.ShopDaoImpl 类下实现接口方法：

```
public class ShopDaoImpl implements IShopDao {
/*
```

(1) 分页查询大类商品。代码如下：

```
*/
@Override
public List<Shop> getShopCategoryPage(int cid, int currentPage, int pageSize)
{ List<Shop> shops = new ArrayList<Shop>();
 String strSql = "select * from shop where cid=? order by pid desc limit ?,?"; //表示查询哪个大类,从哪条记录开始,到哪条记录结束
 return (List<Shop>) DBUtil.executeQuery(strSql, new IResultSetUtil() {
 @Override
 public Object doHandler(ResultSet rs) throws SQLException {
while (rs.next()) {
    Shop shop = new Shop();
    shop.setPid(rs.getInt("pid"));
    shop.setShopname(rs.getString("shopname"));
    shop.setImage1(rs.getString("image1"));
    shop.setPrice(rs.getFloat("price"));
    shops.add(shop);}
    return shops; }
   },cid,(currentPage-1)*pageSize,pageSize);
   }
/*
```

【代码说明】

该方法用来分页查询表 shop 中商品大类的所有商品，以商品大类编号、请求页号和每页记录为参数，首先把每一条查询记录封装到商品实体类 shop 中，然后再把商品实体类 shop 添加到商品集合 shops 中，返回结果为商品集合分页边界类型，即从(currentPage-1)*pageSize 条记录开始，共显示 pageSize 条记录。

(2) 查询每条大类下的商品数目，返回结果为整型。代码如下：

```
*/
@Override
public int getTotalByCid(int cid) {
    String strSql = "select count(*) from shop where cid=? ";
```

```
        Object obj= DBUtil.executeQuery(strSql,cid);
         return Integer.parseInt(obj.toString());
}
/*
```

(3) 按商品名称进行模糊查询。

该方法用来分页模糊查询表 shop 的商品信息，以商品名称、请求页码和每页记录数作为查询参数，返回结果为商品集合分页边界类型。代码如下：

```
*/
@Override
public List<Shop> getShopByShopname(String shopname, int currentPage,
int pageSize) {
List<Shop> shops = new ArrayList<Shop>();
String strSql = "select subcate.sname as sname, shop.* from
shop,subcate where shop.shopname like ? and subcate.sid=shop.sid order
by shop.pid desc limit ?,?";
    return (List<Shop>) DBUtil.executeQuery(strSql, new IResultSetUtil()
{ @Override
    public Object doHandler(ResultSet rs) throws SQLException {
        while (rs.next()) {
            Shop shop = new Shop();
            shop.setPid(rs.getInt("pid"));
            SubCate subCate = new SubCate();
            subCate.setSname(rs.getString("sname"));
            shop.setSubCate(subCate);
            shop.setShopname(rs.getString("shopname"));
            shop.setImage1(rs.getString("image1"));
            shop.setPrice(rs.getFloat("price"));
            shop.setStock(rs.getInt("stock"));
            shops.add(shop); }
         return shops;
         }
     },"%" + shopname + "%",(currentPage - 1) * pageSize,pageSize);
    }
/*
```

【代码说明】

由于是模糊查询，所以在书写查询参数时一定要用"%" + shopname + "%"隔开，由于模糊查询结果不止一条记录，所以要用分页显示。

(4) 按商品名称进行分页模糊查询的总记录数。代码如下：

```
*/
  @Override
 public int getTotalByShopname(String shopname) {
 String strSql = "select count(*) from shop where shopname LIKE ?";
 Object obj= DBUtil.executeQuery(strSql,"%" + shopname + "%");
  return Integer.parseInt(obj.toString()); }
    }
```

(二)商品分类显示子模块业务逻辑层设计

1. 接口类设计

在 com.qzmall.service.IShopService 类下新增以下方法:

```java
public interface IShopService {
    //分页查询大类商品
 public List<Shop> getShopCategoryPage(int cid, int currentPage, int pageSize);
 public int getTotalByCid(int cid);
    //按商品名称分页模糊查询
 public List<Shop> getShopByShopname(String shopname,int currentPage,int pageSize);
 public int getTotalByShopname(String shopname);
    }
```

2. 接口实现类设计

在 com.qzmall.service.impl.ShopServiceImpl 类下实现接口类的方法:

```java
public class ShopServiceImpl implements IShopService {
    @Override  //分页查询大类商品
 public List<Shop> getShopCategoryPage(int cid, int currentPage, int pageSize) {
     IShopDao shopDao=new ShopDaoImpl();
     return shopDao.getShopCategoryPage(cid,currentPage,pageSize);
     }
    @Override
    public int getTotalByCid(int cid) {
     IShopDao shopDao=new ShopDaoImpl();
     return shopDao.getTotalByCid(cid) ; }
    @Override  //按商品名称分页模糊查询
    public List<Shop> getShopByShopname(String shopname, int currentPage, int pageSize) {
     IShopDao shopDao=new ShopDaoImpl();
     return shopDao.getShopByShopname(shopname,currentPage,pageSize);
     }
     @Override //按商品名称分页模糊查询的记录数
    public int getTotalByShopname(String shopname) {
     IShopDao shopDao=new ShopDaoImpl();
     return shopDao.getTotalByShopname(shopname); }
}
```

(三)分页工具类设计

分页工具类放在 com.qzmall.util 包下。

1. 设计全局变量 SHOP_PAGE_SIZE

代码如下:

```java
package com.qzmall.util;
public class Tally {
```

```
public static final int SHOP_PAGE_SIZE = 5;//每页显示商品个数
}
```

2. 建立分页类 Pager

代码如下:

```java
package com.qzmall.util;
public class Pager {
private int currentPage;//当前页
private int totalSize;//总记录数
private int totalPage;//总页数
private boolean hasFirst; //首页
private boolean hasPrevious;//上一页
private boolean hasNext;//下一页
private boolean hasLast;//尾页
public Pager(int currentPage,int totalSize){
    this.currentPage=currentPage;
    this.totalSize=totalSize;
}
public int getCurrentPage() {
    return currentPage;
}
public void setCurrentPage(int currentPage) {
    this.currentPage = currentPage;
}
public void setTotalSize(int totalSize) {
    this.totalSize = totalSize;
}
public int getTotalPage() {
    totalPage=totalSize/Tally.SHOP_PAGE_SIZE;
    if(totalSize%Tally.SHOP_PAGE_SIZE!=0)
    totalPage++;
    return totalPage;
}
public void setTotalPage(int totalPage) {
    this.totalPage = totalPage;
}
public boolean isHasFirst() {
    if(currentPage==1 ){//如果当前页为1,首页链接不能用
        return false;
    }
    return true;
}
public void setHasFirst(boolean hasFirst) {
    this.hasFirst = hasFirst;
}
public boolean isHasPrevious() {
    if(isHasFirst())
    //如果首页链接能用,则可以有上一页链接
        return true;
```

```java
        else
            return false;
    }
    public void setHasPrevious(boolean hasPrevious) {
        this.hasPrevious = hasPrevious;
    }
    public boolean isHasNext() {
        if(isHasLast())
           //如果尾页链接能用,则可以有下一页链接
            return true;
        else
            return false;
    }
    public void setHasNext(boolean hasNext) {
        this.hasNext = hasNext;
    }
    public boolean isHasLast() {
        if(currentPage == getTotalPage())
            //如果当前页为总页数,尾页链接不能用
            return false;
        else
            return true;
    }
    public void setHasLast(boolean hasLast) {
        this.hasLast = hasLast;
    }
}
```

【代码说明】

进行分页需要考虑的几个因素有:当前页(currentPage)、总记录数(totalSize)、总页数(totalPage)和每页显示多少商品记录(SHOP_PAGE_SIZE)。totalSize、totalPage、SHOP_PAGE_SIZE 之间的关系如下:

```
totalPage=totalSize/Tally.SHOP_PAGE_SIZE;
    if(totalSize%Tally.SHOP_PAGE_SIZE!=0)
        totalPage++;
```

(四)商品分类显示子模块控制层设计

1. 大类商品分页显示

在 com.qzmall.servlet.shop 类下建立 servlet:CateShopServlet。具体代码如下:

```java
@WebServlet(name = "CateShopServlet",urlPatterns = "/cateList.do")
public class CateShopServlet extends HttpServlet {
protected void doGet(HttpServletRequest req, HttpServletResponse resp)
throws ServletException, IOException {
    try {
      int currentPage = Integer.parseInt(req.getParameter("currentPage"));
      int cid = Integer.parseInt(req.getParameter("cid"));
      String cname = req.getParameter("cname");
```

```
    IShopService categoryService = new ShopServiceImpl();
    int totalSize = categoryService.getTotalByCid(cid);
    int totalPage = totalSize / Tally.SHOP_PAGE_SIZE;
    if (totalSize % Tally.SHOP_PAGE_SIZE != 0)
            totalPage++;
    if (currentPage < 1) {currentPage = 1;}
    if (currentPage > totalPage){currentPage = totalPage;}
    Pager page = new Pager(currentPage, totalSize);
    List<Shop> shopList = categoryService.getShopCategoryPage(cid,
currentPage, Tally.SHOP_PAGE_SIZE);
    req.setAttribute("shopList", shopList);
    req.setAttribute("page", page);
    String url = req.getRequestURI() + "?" + req.getQueryString();
    System.out.println("url=" + url);
    int index = url.lastIndexOf("&currentPage=");
    if (index != -1) {url = url.substring(0, index);}
        System.out.println("urllast=" + url);
        req.setAttribute("url", url);
        req.setAttribute("cname", cname);
        req.getRequestDispatcher("list.jsp").forward(req, resp);
     } catch (NumberFormatException ex) {
        String msg = "页码传输异常";
      req.setAttribute("msg", msg);
req.getRequestDispatcher("result.jsp").forward(req, resp);     }
   }
}
```

【代码说明】

当用户在 main.jsp 页面，单击商品某一大类时，就会交给 CateShopServlet 类进行处理。

(1) 利用 CateShopServlet 得到页面传来数据。

① 首先通过 req 得到当前页数据 currentPage 和大类编号 cid。

② 通过调用商品业务逻辑对象的 getTotalByCid(cid)得到 totalSize。

③ 通过调用商品业务逻辑对象的 getShopCategoryPage 方法得到 shopList。

④ 通过 req.getRequestURI() + "?" + req.getQueryString()得到 url 地址。

例如：

```
url="/qzmall/subList.do?sid=2&cname=%E6%B4%97%E8%A1%A3%E6%9C%BA&
currentPage= 1"
```

它表示当前执行的查询是"cid=2，currentPage=1"的查询。

如果继续查询下一页或上一页，cid=2 的数据不能丢，currrentPage 这个值是需要变化的，为了达到这个目的，可以把 url 中的 currentPage 的部分去掉，查询时再把它加上，就可以达到查询目的了。

(2) 把得到的有关数据传回页面(list.jsp)。

① 得到分页数据：req.setAttribute("shopList", shopList)。

② 得到分类导航：req.setAttribute("page", page)。
③ 得到截取后的 url 字符串：req.setAttribute("url", url)。
④ 得到分类名称：req.setAttribute("cname", cname)。

(3) 异常处理。

如果输入的当前页不是数字，则提示输入错误。

2. 商品模糊查询功能实现

在 com.qzmall.servlet.shop 类下建立 servlet:QueryShopServlet。具体代码如下：

```java
@WebServlet(name = "QueryShopServlet",urlPatterns = "/queryShop.do")
public class QueryShopServlet extends HttpServlet {
protected void doGet(HttpServletRequest req, HttpServletResponse resp)
throws ServletException, IOException {doPost(req, resp); }
 protected void doPost(HttpServletRequest req, HttpServletResponse resp)
throws ServletException, IOException {
  String shopname = req.getParameter("shopname");
  try {
   int currentPage = Integer.parseInt(req.getParameter("currentPage"));
   IShopService shopService = new ShopServiceImpl();
   int totalSize = shopService.getTotalByShopname(shopname);
   int totalPage=totalSize/Tally.SHOP_PAGE_SIZE;
  if(totalSize%Tally.SHOP_PAGE_SIZE!=0)
            totalPage++;
if (currentPage < 1) { currentPage = 1;}
  if (currentPage > totalPage){ currentPage = totalPage;}
  Pager page = new Pager(currentPage, totalSize);
  List<Shop> shopList = shopService.getShopByShopname(shopname,
currentPage, Tally.SHOP_PAGE_SIZE);
  req.setAttribute("shopList", shopList);
     req.setAttribute("page", page);
     String url = req.getRequestURI() + "?" + req.getQueryString();
     System.out.println("url=" + url);
     int index = url.lastIndexOf("&currentPage=");
     if (index != -1) {url = url.substring(0, index);}
     System.out.println("urllast=" + url);
     req.setAttribute("url", url);
     req.setAttribute("cname", shopname);
     req.getRequestDispatcher("list.jsp").forward(req, resp);
      } catch (NumberFormatException ex) {
       String msg="页码传输异常";
       req.setAttribute("msg",msg);
       req.getRequestDispatcher("result.jsp").forward(req,resp);
      }
   }
}
```

(五)商品分类显示子模块视图层——商品分类页面(list.jsp)设计

代码如下：

```jsp
<%@ taglib prefix="c" uri="http://java.sun.com/jsp/jstl/core" %>
<%@ page contentType="text/html;charset=UTF-8" language="java" %>
<!DOCTYPE html>
<html>
<head>
    <meta charset="UTF-8">
    <meta http-equiv="X-UA-Compatible" content="ie=edge">
    <title>商品列表</title>
    <link type="text/css" rel="stylesheet" href="css/font-awesome.css">
    <link type="text/css" rel="stylesheet" href="css/reset.css">
    <link type="text/css" rel="stylesheet" href="css/list.css">
    <script src="https://libs.baidu.com/jquery/2.1.4/jquery.min.js"></script>
</head>
<body>
<jsp:include page="head.jsp"/>
<div class="crumb">
    <div class="w">
        <div class="crumb-con">
            <a href="index.do" class="link">QZMAIL</a>
            <span></span>
            <span class="link-text">${cname}
            </span>
        </div>
    </div>
</div>
<div class="page-wrap w">
    <!--list容器-->
    <ul class="p-list-con">
        <c:forEach items="${shopList}" var="shop">
        <li class="p-item">
            <div class="p-img-con">
    <a href="detail.do?pid=${shop.pid}" class="link"><img src="<c:url value='img/pic/${shop.image1 }'/>" class="p-img"></a>
            </div>
            <div class="p-price-con">
                <span class="p-price">¥<strong>${shop.price}</strong></span>
            </div>
            <div class="p-name-con">
                <a href="" class="p-name">${shop.shopname}</a>
            </div>
            <div class="p-buy"><a href="cartServlet.do?pid=${shop.pid}&action=add&quantity=1" target="_blank">立即抢购</a></div>
        </li>
        </c:forEach>
    </ul>
    <!--分页容器-->
    <div class="pg-content">
```

```html
        <form class="page-form" method="post" onsubmit="return checkForm();" action="${url}">
            <input type="text" name="currentPage" id="currentPage" class="page" >
            <input type="submit" value="go" >
        </form>
    <c:if test="${page.hasFirst}">
        <li><a href="${url}&currentPage=1">首页</a></li>
    </c:if>
    <c:if test="${page.hasPrevious}">
        <li><a href="${url}&currentPage=
            ${page.currentPage-1}">上一页</a></li>
    </c:if>
    <c:if test="${page.hasNext}">
        <li><a href="${url}&currentPage=
            ${page.currentPage+1}">下一页</a></li>
    </c:if>
    <c:if test="${page.hasLast}">
        <li><a
            href="${url}&currentPage=
    ${page.totalPage}">尾页</a></li>
    </c:if>
     <span class="pg-total">
     当前第 ${page.currentPage} 页, 总共 ${page.totalPage} 页 </span>
     </div>
</div>
<script type="text/javascript">
    function checkForm(){
        var t=$("#currentPage").val();//这个就是我们要判断的值了
        if(isNaN(t)){
            alert("页码不是数字");
            return false;
        }
    }
</script>
<jsp:include page="foot.jsp"/>
</body>
</html>
```

【代码说明】

(1) 本页面接收来自 main.jsp 发出的大类查询、子类查询及模糊查询命令生成的数据。

(2) 利用 JSTL 标签对传来的 shoplist 商品集合进行遍历，得到所要查询的商品信息，利用 JSTL 标签对传来的 page 类进行条件查询操作，<c:if test="${page.hasFirst}">表示如果有首页函数为真，显示首页标签。"<a href="${url}&currentPage=1">首页</a></li>" 中的 url 表示第一次查询生成地址栏地址的字符串去掉 currentPage 后面的字符。这样可以在页码变动的情况下，保证查询条件不会丢。其他以此类推。

(3) 分页容器中还包括一个表单，表单中的文本框是让用户输入页码进行查询。用户输入的页码必须是数字且不能超出页码范围，否则提示出错。

商品分类查询和模糊查询的运行效果分别如图 5-3 和图 5-4 所示。

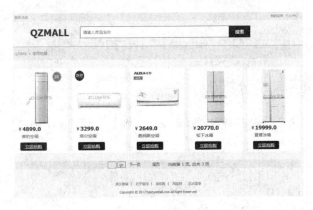

图 5-3 大类商品

图 5-4 商品模糊查询

小提示：Port 8080 is already is use 异常处理

在开发工程中常常会遇到 java.net.BindException: Address already is use:bind 或者 Port 8080 is already is use 等异常，然后 Web Server 不能启动等问题，这是因为已经有其他程序或者进程占用了相应的端口，解决这个问题就要找出到底是哪个程序占用了这个端口，然后关掉那个程序或者取消那个进程就可以了。

解决的步骤如下：
(1) 打开一个命令窗口，执行 netstat-a -o 命令。
(2) 找到占用那个端口的程序，找到它的 pid。
(3) 执行命令 tasklist，通过 pid，找到和 pid 对应的应用程序进程。
(4) 打开任务管理器，终止这个进程。
(5) 重新启动或者运行服务程序，就会发现这个异常没有了。

课堂技能训练

【实训操作内容】商品模糊查询功能实现。
【实训操作步骤】
(1) 持久层设计(重点)。
(2) 业务逻辑层设计。
(3) 控制层设计。
(4) 视图层设计。

第三节 商品详情子模块设计

一、商品详情显示功能分析

商品详情显示包括商品图片轮播显示、基本信息显示、详情信息显示。

图片轮播显示就是把商品表里的商品图片信息转换为图片显示即可，图片轮播功能由 jQuery 实现。

基本信息主要包括商品名称、价格、库存等信息。

详细信息主要包括售后服务、商品配置、用户评价等信息，主要由切换标签来显示。

商品分类显示.flv

二、商品详情显示功能实现

(一)商品详情子模块持久层设计

1. 持久层接口类设计

在 com.qzmall.dao.IShopDao 接口类下新增方法：

```java
public interface IShopDao {
    //查询单个商品详细信息
    public Shop getShopByid(int pid);
}
```

2. 在 com.qzmall.dao.impl.ShopDaoImpl 类下实现上述方法

代码如下：

```java
public class ShopDaoImpl implements IShopDao {
/*
按商品 id 查询商品详细信息
该方法用来分页查询表 sho 中每一条详细信息，以商品 id 作为查询参数，返回结果为商品实体类 shop
*/
@Override
public Shop getShopByid(int pid) {
String strSql = "select  subcate.sid as ssid,shop.* from shop,subcate where shop.sid=subcate.sid and pid=?";
return (Shop) DBUtil.executeQuery(strSql, new IResultSetUtil() {
 @Override
 public Object doHandler(ResultSet rs) throws SQLException {
     if (rs.next()) {
         Shop shop = new Shop();
         shop.setPid(rs.getInt("pid"));
             SubCate subCate = new SubCate();
             subCate.setSid(rs.getInt("ssid"));
```

```
                shop.setSubCate(subCate);
                shop.setShopname(rs.getString("shopname"));
                shop.setShopinfo(rs.getString("shopinfo"));
                shop.setShopdate(rs.getString("shopdate"));
                shop.setStock(rs.getInt("stock"));
                shop.setPrice(rs.getFloat("price"));
                shop.setImage1(rs.getString("image1"));
                shop.setImage2(rs.getString("image2"));
                shop.setImage3(rs.getString("image3"));
                shop.setDescription(rs.getString("description"));
                return shop;
            } else {
                return null;
            }
        }
    },pid); }
}
```

(二)商品详情子模块业务逻辑层设计

1. 接口类设计

在 com.qzmall.service.IShopService 类下新增以下方法：

```
public interface IShopService {
    //查询单个商品详细信息
    public Shop getShopByid(int pid);
}
```

2. 接口实现类设计

在 com.qzmall.service.impl.ShopServiceImpl 类下实现接口类的方法：

```
public class ShopServiceImpl implements IShopService {
    @Override  //查询单个商品详细信息
    public Shop getShopByid(int pid) {
        IShopDao shopDao=new ShopDaoImpl();
        return shopDao.getShopByid(pid);
    }
}
```

(三)商品详情子模块控制层设计——商品详情页功能实现

当用户在 main.jsp 页面单击商品图片时，就会交给 QueryShopServlet 类进行处理，调用商品业务逻辑类方法 getShopByid 生成商品详细信息对象 shop，并回传到商品详细页 (detail.jsp)。代码如下：

```
@WebServlet(name = "DetailServlet",urlPatterns = "/queryShop.do")
public class DetailServlet extends HttpServlet {
    protected void doGet(HttpServletRequest req, HttpServletResponse resp)
    throws ServletException, IOException {
```

```
        int pid=Integer.parseInt(req.getParameter("pid"));
        IShopService shopService=new ShopServiceImpl();
        Shop shop=shopService.getShopByid(pid);
        req.setAttribute("shop",shop);
req.getRequestDispatcher("detail.jsp").forward(req,resp);
    }
}
```

(四)商品详情子模块视图层——商品详情页(detail.jsp)设计

代码如下：

```
<%@ taglib prefix="c" url="http://java.sun.com/jsp/jstl/core" %>
<%@ page contentType="text/html;charset=UTF-8" language="java" %>
<!DOCTYPE html>
<html>
<head>
    <meta charset="utf-8" />
    <meta http-equiv="X-UA-Compatible" content="ie=edge">
    <title>商品详情</title>
    <link type="text/css" rel="stylesheet" href="css/font-awesome.css">
    <link type="text/css" rel="stylesheet" href="css/reset.css">
    <link type="text/css" rel="stylesheet" href="css/basic.css">
    <link type="text/css" rel="stylesheet" href="css/detail.css">
    <script src="https://libs.baidu.com/jquery/2.1.4/jquery.min.js"></script>
    <script type="text/javascript" src="js/detail.js"></script>
</head>
<body>
<jsp:include page="head.jsp"/>
<div class="crumb">
    <div class="w">
        <div class="crumb-con">
            <a href="index.do" class="link">MMail</a>
            <span>></span>
            <span class="link-text">商品详情</span>
        </div>
    </div>
</div>
<div class="w">
    <div class="info-img">
    <div class="main-img">
    <ul class="main-img-info">
      <li><img src="<c:url value='img/pic/${shop.image1}'/>" width="400" height="400" /></li>
      <li> <img src="<c:url value='img/pic/${shop.image2}'/>" width="400" height="400" /></li>
      <li ><img src="<c:url value='img/pic/${shop.image3}'/>" width="400" height="400" /> </li>
```

```html
            </ul>
            <div class="btn btn_l">&lt;</div>
            <div class="btn btn_r">&gt;</div>
        </div>
        <div>
            <ul class="min-img">
                <li ><img src="<c:url value='img/pic/${shop.image1}'/>" width="75" height="75" /></li>
                <li ><img src="<c:url value='img/pic/${shop.image2}'/>" width="75" height="75" alt=""/></li>
                <li><img src="<c:url value='img/pic/${shop.image3}'/>" width="75" height="75" alt=""/></li>
            </ul>
        </div>
    </div>
    <div class="p-info-con">
        <h1 class="p-name">${shop.shopname}</h1>
        <p class="p-subtitle">${shop.shopinfo}</p>
        <div class="p-info-item p-price-con">
            <span class="label">价    格:    </span>
            <span class="info">¥${shop.price}</span>
        </div>
        <div class="p-info-item">
            <span class="label">库    存:</span>
            <span class="info">${shop.stock}</span>
        </div>
        <div class="p-info-item">
            <span class="label">上架时间:</span>
            <span class="info adddate">${shop.shopdate}</span>
        </div>
        <div class="p-buy"><a href="cartServlet.do?pid=${shop.pid}&action=add&quantity=1" target="_blank">立即抢购</a></div>
    </div>
</div>
</div>
<div class="w">
    <div class="detail-wrap">
        <ul class="tab-list">
            <li class="active">商品详情</li>
            <li >规格参数</li>
            <li >包装与售后</li>
            <li >用户评价</li>
        </ul>
        <div class="content detail-con">
            ${shop.description}
        </div>
        <div class="detail-con">
            规格参数
        </div>
        <div class="detail-con">
```

```
            包装与售后
        </div>
        <div class="detail-con">
            用户评价
        </div>
    </div>
</div>
<jsp:include page="foot.jsp"/>
</body>
</html>
```

【代码说明】

(1) 本页面主要用来显示商品的详细信息，接收来自 DetailServlet 传来的 shop 实体类对象，由于只显示一条商品信息，所以直接用 EL 标签来进行解析就可以了。

(2) 本页面左边部分显示的商品轮播图，用 EL 标签显示的 3 张商品图片来完成。

(3) 右上部分显示商品基本信息，如商品名称、价格、库存等。

(4) 右下部分的标签切换，主要用来显示商品详细信息，如规格参数、包装售后、用户评价，这样更能节省空间。

(5) 轮播图和标签切换 jQuery 代码请参照第四章 jQuery 应用部分。

商品详情页运行效果如图 5-5 所示。

图 5-5　商品详情页

小提示：request.getAttribute()和 request.getParameter()的区别

(1) getAttribute()有对应的 setAttribute()方法，而 getParameter()没有对应的 setParameter()方法。

(2) getParameter()获取到的值只能是字符串，不可以是对象，而 getAttribute()获取到的值是 Object 类型的。

(3) 通过 form 表单或者 url 来向另一个页面或者 servlet 传递参数的时候需要用 getParameter 获取值；getAttribute 只能获取 setAttribute 的值。

(4) setAttribute()是应用服务器把这个对象放到该页面所对应的一块内存当中,当页面服务器重定向到另一个页面的时候,应用服务器会把这块内存复制到另一个页面对应的内存当中。通过getAttribute()可以取得存储下的值,当然这种方法可以用来传对象。

用session也是一样的道理,只是request和session的生命周期不一样而已。

课堂技能训练:

【实训操作内容】商品详情显示功能实现。

【实训操作步骤】

(1) 持久层设计。

(2) 业务逻辑层设计。

(3) 控制层设计。

(4) 视图层设计。

课 后 训 练

一、选择题

1. 在JSP小脚本中,使用下面(   )语句可以正确重定向至另外一个页面。

    A. req.getRequestDispatcher("index.jsp").forward(req,resp)

    B. resp.sendRedirect("ndex.jsp")

    C. req.sendRedirect("index.jsp")

    D. resp.sendRedirect()

2.
```
public void doFilter(ServletRequest req, ServletResponse resp,
FilterChain chain) throws ServletException, IOException {
chain.doFilter(req, resp);
RequestDispatcher dis;
_____;
dis.forward(req,resp);}
```

那么能够正确填写在横线处的选项是(   )。

    A. dis=new RequestDispatcher("error.jsp");

    B. dis=request.getRequestDispatcher("error.jsp");

    C. dis=response.getRequestDispatcher("error.jsp");

    D. dis=request.RequestDispatcher("error.jsp");

3.
```
@WebServlet(name = "sdServlet",urlPatterns = "/sd.do")
public class sdServlet extends HttpServlet {
    protected void doGet(HttpServletRequest req, HttpServletResponse resp)
```

```
    throws ServletException, IOException {
        ServletContext sc=req.getServletContext();
        sc.setAttribute("name","小明");
        resp.sendRedirect("index.jsp");
    }
}
```

以下是 index.jsp:

```
<% String name=_____;
   out.print(name);
%>
```

index.jsp 页面中获取 name 数据的正确写法是(   )。

    A. (String)request.getAttribute("name")

    B. (String)application.getAttribute("name")

    C. request.getAttribute("name")

    D. application.getAttribute("name")

4. (   )是同应答相关的 HttpServletResponse 类的一个对象,它封装了服务器对客户端的响应,然后被送到客户端以响应客户的请求。

    A. request   B. pageContext

    C. response   D. out

5. 要在 JSP 中使用 ArrayList,正确的做法是(   )。

    A. <%@ page import ="java.util.ArrayList" %>

    B. <% import java.util.ArrayList %>

    C. <%@ import "java.util.ArrayList" %>

    D. <%@ page package="java.util.ArrayList" %>

二、实际操作题

进行用户模块的持久层、业务逻辑层、控制层、视图层方法设计,实现子类商品分页显示功能。

第五章 商品显示模块设计.pptx

第五章 习题答案.docx

第六章
用户模块设计

> **知识能力目标**
>
> 1. 学会利用 Servlet 读取表单数据和 session 对象数据。
> 2. 了解用户注册的 SQL 语句、步骤、完成要求及应注意的问题。
> 3. 了解用户登录的 SQL 语句、步骤、完成要求及应注意的问题。
> 4. 了解用户积分的奖励办法。
> 5. 掌握用户修改个人密码的流程设计。
> 6. 能够完成用户注册各项要求的编码设计。
> 7. 能够完成用户登录各项要求的编码设计。
> 8. 能够完成用户积分模块的编码设计。
> 9. 能够完成用户修改密码的编码设计。

问题提示

电子商务平台几乎家家都有积分，积分是商家为了刺激消费者消费，是一种变相营销的手段。积分一般是用户在商城购物时由商城根据用户购物的多少给用户的奖励。积分有很多用途，不仅在购买商城商品时可用于抵现(100 商城积分=1 元)，更可在商城积分频道以及相关活动中直接兑换超值商品。淘宝商城积分可以在积分兑换频道参与积分秒杀、积分兑换、积分加钱购等多项活动，并且在购物时可以抵现。

问题：电子商城中积分获取的规则是什么？用户怎么查看自己的积分？

第一节 用户注册子模块设计

一、用户注册的流程

用户注册功能对于电子商务网站的重要性毋庸置疑，注册模块是设计中比较复杂的一个模块。用户注册的基本流程：输入用户信息；对用户输入的信息进行前台校验(用户名校验、验证码校验、密码、手机号、注册时间)；控制层得到用户输入信息；校验用户输入信息；系统将会员信息录入数据库；给注册用户一定的积分奖励。

二、基础知识

(一)利用 Servlet 读取表单数据

(1) request.getParameter(String name)：此方法用来获取表单参数的值。

(2) request.getParameterValues(String name)：如果参数出现一次以上，则调用该方法，并返回多个值，例如复选框。

(3) request.getParameterNames()：如果想要得到当前请求中的所有参数的完整列，则调用该方法。

(二)利用 Servlet 读取 session 对象数据

(1) HttpSession session = request.getSession()：获取 session。
(2) User user=(User)session.getAttribute(String name)：读取 session 对象。
(3) Session.setAttribute(String name,Object obj)：给 session 对象赋值。
(4) session.removeAttribute(String name)：删除当前 session。
(5) session. invalidate()：注销当前的 session。

用户更改信息的流程：信息有误—提交修改信息—后台检验—调用业务逻辑层方法更新用户信息。

三、用户注册功能实现

(一)用户注册子模块持久层设计

1. 接口类设计

在 com.qzmall.dao 包下，新建接口类 IUserDao：

```
public interface IUserDao {
    //注册用户信息接口
    public boolean addUser(User user);
    //用户积分接口
    public boolean addPoint(Point point); }
```

2. 接口实现类设计

在 com.qzmall.dao.impl 类下创建接口实现类 UserDaoImpl：

```
public class UserDaoImpl implements IUserDao {
/*
```

(1) 用户注册。

该方法用来实现往用户表插入一条记录，以 User 对象为参数，返回结果为布尔类型，通过调用数据库操作类中的方法 DBUtil.executeUpdate 执行更新 SQL 语句，如果影响记录数大于 0，返回 true，否则返回 false。主要代码如下：

```
*/
@Override
public boolean addUser(User user) {
String strSql = "insert into user
(username,password,birthday,adddate,question,answer,point,truename,phone,
postcode,address) values (?,?,?,?,?,?,?,?,?,?,?) ";
    return DBUtil.executeUpdate(strSql, user.getUsername(),user.getPassword(),
user.getBirthday(),user.getAdddate(),user.getQuestion(),user.getAnswer(),
user.getPoint(),user.getTruename(),user.getPhone(),user.getPostcode(),user.
getAddress()) > 0;
}
/*
```

(2) 用户积分维护。

该方法的主要作用是当用户注册成功后或购物成功后，随之往用户积分表添加一条记录，以记录用户积分信息。代码如下：

```
*/
@Override
public boolean addPoint(Point point) {
    String strSql = "insert into point (username,orderid,mypoint,pointtime,description) values (?,?,?,?,?)";
    return DBUtil.executeUpdate(strSql, point.getUsername(), point.getOrderid(), point.getMypoint(), point.getPointtime(), point.getDescription()) > 0;
}
```

(二)用户注册子模块业务逻辑层设计

1. 接口类设计

在 com.qzmall.service 包下新建 IUserService 类，接口类设计的方法与持久层相同，这里不再赘述。

2. 接口实现类设计

在 com.qzmall.service.impl 包下建立接口实现 UserServiceImpl 类。具体代码如下：

```
public class UserServiceImpl implements IUserService {
    @Override
    public boolean addUser(User user) {
        IUserDao userDao=new UserDaoImpl();
        return  userDao.addUser(user);  }
    @Override
    public boolean addPoint(Point point) {
        IUserDao userDao=new UserDaoImpl();
        return  userDao.addPoint(point);  }
}
```

(三)用户注册子模块控制层设计

在 com.qzmall.servlet.user 包下新建 AddUserServlet 类。具体代码如下：

```
@WebServlet(name = "AddUserServlet",urlPatterns = "/addUserServlet.do")
public class AddUserServlet extends HttpServlet {
    private static final long serialVersionUID = 1L;
    //用来表明类的不同版本间的兼容性
    @Override
    protected void doPost(HttpServletRequest req, HttpServletResponse resp) throws ServletException, IOException{
        Point point=new Point();
        IUserService userService=new UserServiceImpl();
        String username=req.getParameter("username");
```

```java
            if (userService.getUserByUsername(username)){
                String result="用户名错误";
req.setAttribute("erroruser",result);
req.getRequestDispatcher("register.jsp").forward(req,resp);}
            String password=req.getParameter("password");
            String truename=req.getParameter("truename");
            String phone=req.getParameter("phone");
            String address=req.getParameter("address");
            String birthday=req.getParameter("birthday");
            String postcode=req.getParameter("postcode");
            String code=req.getParameter("yzm");
            int question=Integer.parseInt(req.getParameter("question"));
            String answer=req.getParameter("answer");
            point.setOrderid("");
            point.setUsername(username);
            point.setDescription("注册奖励积分");
            point.setMypoint(15);
            point.setPointtime(StringHelper.getCurrentFormatDate());
            User user=new User();
            user.setUsername(username);
            user.setPassword(StringHelper.MD5(password));
            user.setAddress(address);
            user.setPhone(phone);
            user.setTruename(truename);
            user.setQuestion(question);
            user.setAnswer(answer);
            user.setPostcode(postcode);
            user.setBirthday(birthday);
            user.setAdddate(StringHelper.getCurrentFormatDate());
            user.setPoint(0);
            if (userService.addUser(user)) {
                userService.addPoint(point);
                String result = "恭喜你,注册成功!";
                req.setAttribute("msg", result);
                req.getRequestDispatcher("result.jsp").forward(req, resp);
            } else {
                String result = "注册失败!";
                req.setAttribute("msg", result);
                req.getRequestDispatcher("result.jsp").forward(req, resp);
            }
        }
```

【代码说明】

当用户在用户注册页面(register.jsp)输入用户信息后,单击"注册"按钮,就会交给AddUserServlet 进行处理,代码首先判断用户输入的用户名是否合法,如果输入正确,执行插入语句,注册成功之后,奖励用户 15 积分,并添加到积分记录表(point)中。

小提示:什么是 ORM

对象关系映射(Object Relational Mapping,ORM)是一种程序技术,用于实现面向对象编程语言里不同类型系统的数据之间的转换。从效果上说，它其实是创建了一个可在编程语言里使用的"虚拟对象数据库"。

面向对象是在软件工程基本原则(如耦合、聚合、封装)的基础上发展起来的，而关系数据库则是从数学理论发展而来的，两套理论存在显著的区别。为了解决这个不匹配的现象，对象关系映射技术应运而生。

对象关系映射提供了概念性的、易于理解的模型化数据的方法。ORM 方法论基于三个核心原则：简单——以最基本的形式建模数据。传达性——数据库结构被任何人都能理解的语言文档化。精确性——基于数据模型创建正确标准化的结构。典型地，建模者通过收集来自那些熟悉应用程序但不熟悉数据建模的人的信息开发信息模型。建模者必须能够用非技术企业专家可以理解的术语在概念层次上与数据结构进行通信。建模者也必须能以简单的单元分析信息，对样本数据进行处理。ORM 专门被设计为改进这种联系。

简单地说，ORM 相当于中继数据。具体到产品上，例如 ADO.NET Entity Framework。DLINQ 中实体类的属性[Table]就算是一种中继数据。

(四)用户注册子模块视图层——注册网页(register.jsp)设计

代码如下：

```
<%@ page contentType="text/html;charset=UTF-8" language="java" %>
<!DOCTYPE html>
<html>
<head>
    <meta charset="utf-8" />
    <meta http-equiv="X-UA-Compatible" content="ie=edge">
    <title>会员注册</title>
    <link type="text/css" rel="stylesheet" href="css/font-awesome.css">
    <link type="text/css" rel="stylesheet" href="css/font-awesome.min.css">
    <link type="text/css" rel="stylesheet" href="css/reset.css">
    <link type="text/css" rel="stylesheet" href="css/register.css">
    <script src="https://libs.baidu.com/jquery/2.1.4/jquery.min.js"></script>
    <script type="text/javascript" src="js/My97DatePicker/WdatePicker.js"></script>
    <script type="text/javascript" src="js/register.js"></script>
</head>
<body>
<jsp:include page="head.jsp"/>
<div class="page-wrap">
    <div class="w">
        <div class="user-con">
            <div class="user-title">用户注册</div>
            <form class="form-horizontal" role="form" id="registerForm" action="addUserServlet.do"
                method="post" novalidate="novalidate" onsubmit="return checkForm();" > <!--onsubmit="return checkForm();" -->
                <div class="user-box">
                    <div class="user-item">
                        <label class="user-label" for="username">
```

```html
                    <i class="fa fa-user"></i>
                </label>
        <input class="user-content" id="username" placeholder="请输入用户名" name="username" autocomplete="off"
        onblur="checkUsername();" onfocus="checkHideUsername();" />
    <span id="span1" >用户名</span>
                </div>
                <div class="user-item">
                    <label class="user-label" for="password1">
                        <i class="fa fa-lock"></i>
                    </label>
            <input class="user-content" id="password1" name="password" placeholder="请输入密码" autocomplete="off" type="password">
                </div>
                <div class="user-item">
                    <label class="user-label" for="password2">
                        <i class="fa fa-lock"></i>
                    </label>
            <input class="user-content" id="password2" name="password2" placeholder="请再次输入密码" autocomplete="off" type="password">
                </div>
                <div class="user-item">
                    <label class="user-label" for="truename">
                        <i class="fa fa-user-plus"></i>
                    </label>
            <input class="user-content" id="truename" name="truename" placeholder="请输入真实姓名" autocomplete="off"> </div>
                <div class="user-item">
                    <label class="user-label" for="phone">
                        <i class="fa fa-phone"></i>
                    </label>
<input class="user-content" id="phone" name="phone"  placeholder="请输入手机号" autocomplete="off">
                </div>
                <div class="user-item">
                  <label class="label-quest" >
                      <i class="fa fa-question"></i>
                  </label>
            <select name="question" id="question" class="user-quest">
                <option value="" selected>请输入密码问题</option>
                <option value="1">你的中学名字</option>
                <option value="2">你的母亲名字</option>
                <option value="3">你喜欢的人物是谁</option>
                 <option value="4">你父亲的生日</option>
                    </select>
                </div>
                <div class="user-item">
                    <label class="user-label" for="answer">
                        <i class="fa fa-key"></i>
                    </label>
 <input class="user-content" id="answer" name="answer" placeholder="请输入密码提示问题答案" autocomplete="off">
                </div>
```

```html
                <div class="user-item">
                    <label class="user-label" for="postcode">
                        <i class="fa fa-barcode"></i>
                    </label>
    <input class="user-content" id="postcode" name="postcode" placeholder="请输入邮编" autocomplete="off">
                </div>
                <div class="user-item">
                    <label class="user-label" for="address">
                        <i class="fa fa-home"></i>
                    </label>
    <input class="user-content" id="address" name="address" placeholder="请输入邮寄地址" autocomplete="off"></div>
            <div class="user-item">
                <label class="user-label" for="birthday">
                    <i class="fa fa-birthday-cake"></i>
                </label>
                <input class="user-content" id="birthday" name="birthday" placeholder="请输入出生日期" autocomplete="off"
                    onclick="WdatePicker()";> </div>
        <button type="submit" class=" btn btn-submit">注册</button>
</div>
            </form>
            <div class="link-item">
             <a href="login.jsp" class="link" >已有账号，去登录</a>
            </div>
        </div>
    </div>
</div>
<jsp:include page="foot.jsp"/>
</body>
</html>
```

注册页面的运行效果如图 6-1 所示。

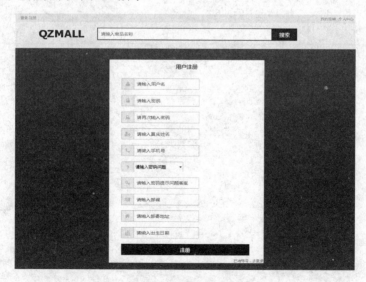

图 6-1 用户注册页面

【代码说明】

(1) onclick="WdatePicker()"函数的作用是当用户单击生日输入框时出现日期控件，这个日期控件是用 jQuery 编写的，使用时必须先引用。代码如下：

```
<script type="text/javascript" src="js/My97DatePicker/WdatePicker.js"></script>
```

(2) checkForm()函数的作用是当用户单击"提交"按钮时，会先在前台对用户输入的各种信息进行合法性验证，比如：用户名只能为英文或数字，且不能以数字打头，长度为 4～20 个字符，密码长度不小于 6 位，另外还对手机号、邮编、邮寄地址进行了验证，当输入各种信息后，单击"注册"按钮，出现注册成功提示信息，如图 6-2 所示，表明注册成功。

图 6-2　注册成功

课堂技能训练：

【实训操作内容】完成用户注册功能。

【实训操作步骤】

(1) 持久层设计(重点)。

(2) 业务逻辑层设计。

(3) 控制层设计(重点)。

【知识拓展】

本节只是简单实现了用户注册的功能，大多数网站当中，对用户注册信息的要求非常严格，由此实现起来也非常复杂。比如：注册时需要登录邮箱进行验证，需要输入手机验证码，需要输入用户头像和身份证，需要输入个人爱好等多项信息，不但对输入信息进行前台验证，还要进行后台验证。有兴趣的读者可以在完成简单功能的基础上，再尝试完成以上这些功能。

第二节　用户登录子模块设计

一、用户登录基本流程

Web 上的用户登录功能是最基本的功能，网页中用户的登录作用和其他任何方面的登录作用相同，这里只是一个身份认证机制。比如修改用户信息，进行购物，没有登录的用户是不能进行这些操作的。

用户登录流程：输入用户名和密码；前台检验；控制层得到用户输入的信息并调用业务逻辑类登录方法实现登录功能；登录成功后，首页应显示登录用户名，并可以退出登录。

二、用户登录功能的实现

(一)用户登录子模块持久层设计

1. 接口类设计

在 com.qzmall.dao.IUserDao 类下新增以下方法：

```
public interface IUserDao {
    //用户登录的接口
    //return (1:用户名错误；2:密码错误；3:正确)
    public User login(User user);
}
```

2. 持久层接口实现类方法设计

在 com.qzmall.dao.impl.UserDaoImpl 类下实现接口方法：

```
public class UserDaoImpl implements IUserDao {
/*
```

该方法用来实现用户登录功能，以 User 对象为参数，通过调用数据库操作类中的方法 DBUtil.executeQuery 执行查询 SQL 语句，如果记录数大于 0，返回结果为 User 对象，主要代码如下：

```
*/
    @Override
    public User login(User user) {
        String strSql = "select * from user where username=? and password=?";
        return (User)DBUtil.executeQuery(strSql, new IResultSetUtil()
{ @Override
        public Object doHandler(ResultSet rs) throws SQLException {
            if (rs.next()) {
                User user1 = new User();
                user1.setUsername(rs.getString("username"));
                user1.setTruename(rs.getString("truename"));
                return user1; }
            return null;}
            },user.getUsername(),user.getPassword()); }
        }
```

(二)用户登录子模块业务逻辑层设计

1. 接口类方法设计

接口类方法放在 com.qzmall.service.IUserService 类下，接口类设计的方法与持久层相同，这里不再赘述。

2. 接口实现类方法设计

接口实现类方法放在 com.qzmall.service.impl.UserServiceImpl 类下，具体代码如下：

```java
public class UserServiceImpl implements IUserService {
   @Override
   /*
    登录的实现方法
    return (1:用户名错误；2：密码错误；3：正确)
   */
   public User login(User user) {
       IUserDao userDao=new UserDaoImpl();
       return  userDao.login(user);
   }
}
```

(三)验证码类(com.qzmall.util.yzm)设计

代码如下：

```java
package com.qzmall.util;
public class yzm {
  public static  String drawImageVerificate(HttpServletResponse resp){   //定义验证码的宽度和高度
       int width = 100,height = 30;
       //在内存中创建图片
       BufferedImage image = new BufferedImage(width,height,BufferedImage.TYPE_INT_RGB);
       //创建图片的上下文
       Graphics2D g = image.createGraphics();
       //产生随机对象，此随机对象主要用于算术表达式的数字
       Random random = new Random();
       //设置背景
       g.setColor(getRandomColor(240,250));
       //设置字体
       g.setFont(new Font("微软雅黑", Font.PLAIN,22));
       //开始绘制
       g.fillRect(0,0,width,height);
       //干扰线的绘制，绘制线条到图片中
       g.setColor(getRandomColor(180,230));
       for(int i=0;i<10;i++){
           int x = random.nextInt(width);
           int y = random.nextInt(height);
           int x1 = random.nextInt(60);
           int y1 = random.nextInt(60);
           g.drawLine(x,y,x1,y1);
       }
       //开始进行对算术验证码表达式的拼接
       int num1 = (int)(Math.random()*10 + 1);
       int num2 = (int)(Math.random()*10 + 1);
```

```java
    int fuhao = random.nextInt(3);//产生一个[0, 2]之间的随机整数
     //记录符号
     String fuhaostr = null;
     int result = 0;
     switch (fuhao){
         case 0 : fuhaostr = "+";result = num1 + num2;break;
         case 1: fuhaostr = "-";result = num1 - num2;break;
         case 2 : fuhaostr = "*";result = num1 * num2;break;
         //case 3 : fuhaostr = "/";result = num1 / num2;break;
     }
     //拼接算术表达式，用户图片显示
     String calc = num1 + " " + fuhaostr +" "+ num2 +" = ?";
     //设置随机颜色
     g.setColor(new Color(20+random.nextInt(110),20+random.nextInt(110),20+random.nextInt(110)));
     //绘制表达式
     g.drawString(calc,5,25);
     //结束绘制
     g.dispose();
     try {
         //输出图片到页面
         ImageIO.write(image,"JPEG",resp.getOutputStream());
         return String.valueOf(result);
     }catch (Exception ex){
         ex.printStackTrace();
         return null;
     }
  }
  /*
   * 范围随机颜色
   */
  public static Color getRandomColor(int fc,int bc){
     //利用随机数
     Random random  = new Random();
     //随机颜色，了解颜色-Color(red,green,blue).rgb 三原色 0-255
     if(fc>255)fc = 255;
     if(bc>255)bc = 255;
     int r = fc+random.nextInt(bc-fc);
     int g = fc+random.nextInt(bc-fc);
     int b = fc+random.nextInt(bc-fc);
     return new Color(r,g,b);
  }
}
```

小提示：验证码的作用

验证码可以有效防止某个黑客对某一个特定注册用户用特定程序暴力破解方式进行不断的登录尝试，实际上使用验证码是现在很多网站通行的方式(比如招商银行的网上个人银行、百度社区)，利用比较简易的方式实现了这个功能。

(四)用户登录子模块控制层——用户登录功能设计

在 com.qzmall.servlet.user 包下新建 LoginServlet 类,主要代码如下:

```java
@WebServlet(name = "LoginServlet",urlPatterns ="/login.do" )
public class LoginServlet extends HttpServlet {
    private static final long serialVersionUID = 1L;
    //用来表明类的不同版本间的兼容性
    @Override
    protected void doPost(HttpServletRequest req, HttpServletResponse resp)  throws ServletException, IOException {
        HttpSession session = req.getSession();
        String result="";
        User user1=new User();
            //获取用户传递过来的验证码、用户名
        String code = req.getParameter("code");
          user1.setUsername(req.getParameter("username"));
            //把获取到的用户的密码用MD5方法进行加密
user1.setPassword(StringHelper.MD5(req.getParameter("password")));
            //获取验证码框架产生的验证码(会话中存储的验证码)
        String sessionCode = (String)session.getAttribute("result");
        if(code!=null&sessionCode!=null) {
            //如果用户输入的验证码和产生在服务器端的验证码一致,那么就告诉用户输入正确
            if (code.equalsIgnoreCase(sessionCode)) {
                //调用业务逻辑处理类
                IUserService userService = new UserServiceImpl();
                //接收返回的结果
              User user=userService.login(user1);
                if (user!=null) {
                    //用户名和密码都正确
                    //把用户名存入到session中
                    req.getSession().setAttribute("user", user);
                    session.setMaxInactiveInterval(3600);
                    result="3";
                }
                else{
                    result="2";
                }
            } else {
                result="4";
            }
        }
        resp.getWriter().print(result);
        PrintWriter out = resp.getWriter();
    }
}
```

【代码说明】

当用户在登录页面(login.jsp)输入用户名和密码及验证码之后,就会交给 LoginServlet

进行处理。如果验证码错误，则返回 4，如果用户名、密码正确则返回 3，如果用户名、密码错误则返回 2。

(五)用户登录功能视图层设计

1. 用户登录页面(login.jsp)设计

代码如下：

```jsp
<%@ taglib prefix="c" url="http://java.sun.com/jsp/jstl/core" %>
<%@ page contentType="text/html;charset=UTF-8" language="java" %>
<!DOCTYPE html>
<html>
<head>
    <meta charset="UTF-8">
    <meta http-equiv="X-UA-Compatible" content="ie=edge">
    <title>秦职电商平台</title>
    <link type="text/css" rel="stylesheet" href="css/font-awesome.css">
    <link type="text/css" rel="stylesheet" href="css/reset.css">
    <link type="text/css" rel="stylesheet" href="css/login.css">
</head>
<body>
<jsp:include page="head.jsp"/>
<div class="page-wrap">
    <div class="w">
        <div class="user-con">
            <div class="user-title">用户登录</div>
            <p class="err-msg">${msg}</p>
            <div class="user-box">
                <div class="error-item">
                    <i class="fa fa-minus-circle error-icon"></i>
                    <p class="err-msg">${msg}</p>
                </div>
                <div class="user-item">
                    <label class="user-label" for="username">
                        <i class="fa fa-user"></i>
                    </label>
    <input class="user-content" id="username" placeholder="请输入用户名" autocomplete="off">
                </div>
                <div class="user-item">
                    <label class="user-label" for="password">
                        <i class="fa fa-lock"></i>
                    </label>
      <input class="user-content" id="password" placeholder="请输入密码" autocomplete="off" type="password">
                </div>
                <div class="user-item">
                    <label class="label-yzm" for="yzm">
                        <i class="fa fa-key"></i>
                    </label>
```

```
        <input class="user-yzm" id="yzm" placeholder="请输入验证码"
autocomplete="off">
        <img class="yzm-img" src="code.jsp" alt="" id="code"
onclick="javascript:changeCode();" >
            </div>
        <input type="submit"  class="btn btn-submit" value="登录">
            </div>
        </div>
    </div>
</div>
<jsp:include page="foot.jsp"/>
</body>
<script type="text/javascript" src="js/jquery-1.11.2.min.js"></script>
<script type="text/javascript" src="js/login.js"></script>
<script>
<!-- 刷新验证码-->
 function changeCode() {
 document.getElementById("code").src = "code.jsp?d="+new
Date().getTime(); }
</script>
</html>
```

2. login.js 代码设计

代码如下：

```
$(function() {
    $(".btn-submit").click(function() {
        var username = $("#username");
        var  code=$("#yzm");
        var password = $("#password");
        if (username.val() == "") {
            alert("用户名不能为空");
            username.focus();
            return false; }
        if (password.val() == "") {
            alert("密码不能为空");
            password.focus();
            return false; }
        if (code.val() == "") {
            alert("验证码不能为空");
             code.focus();
            return false; }
        $.ajax({
            async:false,
            cache:false,
            url : "login.do",
            type : "POST",
            data : {
                "username" : username.val(),
                "password" : password.val(),
```

```
                    "code":code.val(),
                    "time" : new Date().getTime()},
            dataType : "json",
            success : function(result) {
              if (result == 2) {
              $(".error-item").css('display','block')
              $(".error-item .err-msg").html("密码或用户名错误");
                  return false;
                } else if (result == 3) {
                    //正确
                    location.href ='index.do';}
              else if (result ==4 ) {
                $(".error-item").css('display','block')
                $(".error-item .err-msg").html("验证码输入错误");
                    return false; }
              },
            error : function(er) {
                console.log(er);
            }
        });
    });
});
```

【代码说明】

首先对用户输入的用户名、密码、验证码进行非空校验，然后以异步请求方式通过调用 login.do 进行登录验证，验证完成后，返回结果信息，如果返回 3，表明用户名和密码输入正确，可以登录首页。

登录页面的运行效果如图 6-3 所示。

图 6-3 用户登录页面

3. 验证码(Code.jsp)页面设计

代码如下：

```
<%@ page import="com.qzmall.util.yzm" %>
<%@ page contentType="text/html;charset=UTF-8" language="java" %>
```

```
<%
/*目的是清空浏览器缓存，因为浏览器会对网站的资源文件和图像进行记忆存储，如果被浏览器加
载过的图片就记忆起来，记忆以后文件就不会和服务器再交互，如果我们验证不清空的话可能会造
成一个问题：验证刷新以后没有效果。*/
    response.setHeader("pragma","no-cache");
    response.setHeader("cache-control","no-cache");
    response.setHeader("expires","0");
    //调用编写的生成验证码的工具
    String result = yzm.drawImageVerificate(response);
    session.setAttribute("result",result);
    //解决getOutputStream异常问题
    // out.clear();
    // out = pageContext.pushBody();
%>
```

课堂技能训练：

【实训操作内容】 用户登录功能实现。

【实训操作步骤】

(1) 持久层设计。

(2) 业务逻辑层设计。

(3) 控制层设计。

(4) 视图层设计。

要求：用异步请求方式判断验证码、用户名和密码输入是否正确。

第三节 用户个人中心子模块设计

一、个人中心模块主要功能和基本流程

用户个人中心主要功能是显示和修改用户基本信息、修改个人密码、查询用户积分。

显示用户个人信息的基本流程：用户登录个人中心；显示个人注册信息。

修改个人信息的基本流程：用户登录个人中心；输入个人信息；前台检验密码输入是否有误；提交修改信息；后台检验；调用业务逻辑层方法更新用户信息。

二、用户个人信息维护功能实现

(一)用户个人信息子模块持久层设计

1. 接口类设计

在 com.qzmall.dao.IUserDao 类下新增以下方法：

```
public interface IUserDao {
    //查询用户信息接口
    public User getUserByUser(String username);
```

```
    //更新用户信息接口
    public boolean saveUserInfo(User user);
    }
```

2. 持久层接口实现类方法设计

在 com.qzmall.dao.impl.UserDaoImpl 类下实现接口方法：

```
public class UserDaoImpl implements IUserDao {
 /*
```

(1) 显示用户个人信息。

该方法用来实现查询表 user 记录的功能，以用户名作为参数进行查询，查询结果封装到 User 对象，主要通过调用 DBUtil.executeQuery 查询方法执行查询语句。代码如下：

```
 */
    @Override
    public User getUserByUser(final String username) {
     String strSql = "select * from user where username=?";
     return (User)DBUtil.executeQuery(strSql, new IResultSetUtil()
{ public Object doHandler(ResultSet rs) throws SQLException
     {if (rs.next()) {User user = new User();
        user.setUsername(rs.getString("username"));
        user.setPassword(rs.getString("password"));
        user.setTruename(rs.getString("truename"));
        user.setAnswer(rs.getString("answer"));
        user.setQuestion(rs.getInt("question"));
        user.setBirthday(rs.getString("birthday"));
        user.setPoint(rs.getInt("point"));
        user.setPhone(rs.getString("phone"));
        user.setAddress(rs.getString("address"));
        user.setPostcode(rs.getString("postcode"));
        user.setAdddate(rs.getString("adddate"));
         return user; }
         return null;}
         } ,username);}
 /*
```

(2) 修改用户信息。

该方法用来实现修改用户表 user 中用户记录的功能，以 User 对象为参数，返回结果为布尔类型，通过调用数据库操作类中的方法 DBUtil.executeUpdate 执行更新 SQL 语句，如果记录数大于 0，返回 true，否则返回 false。主要代码如下：

```
 */
@Override
public boolean saveUserInfo(User user) {
String strSql = "update user set truename=?,
birthday=?,question=?,answer=?,phone=?,postcode=?,addre ss=? where username=?";
```

```
    return 
DBUtil.executeUpdate(strSql,user.getTruename(),user.getBirthday(),user.
getQuestion(),user.getAnswer(),user.getPhone(),user.getPostcode(),user.
getAddress(),user.getUsername())>0 ; }
 }
```

(二)用户个人信息子模块业务逻辑层设计

1. 接口类方法设计

接口类方法放在 com.qzmall.service.IUserService 类下,接口类设计的方法与持久层相同,这里不再赘述。

2. 接口实现类方法设计

接口实现类方法放在 com.qzmall.service.impl.UserServiceImpl 类下,具体代码如下:

```
public class UserServiceImpl implements IUserService {
    //根据用户账号查询用户信息
    @Override
    public User getUserByUser(String username ) {
        IUserDao userDao=new UserDaoImpl();
        return userDao.getUserByUser(username);
    }
//根据用户账号更新用户信息
    @Override
    public boolean saveUserInfo(User user) {
        IUserDao userDao=new UserDaoImpl();
        return userDao.saveUserInfo(user); }
}
```

(三)用户个人信息子模块控制层方法设计

1. 查询用户信息

在 com.qzmall.servlet.user 包下新建 UserInfoServlet 类,主要代码如下:

```
@WebServlet(name = "UserInfoServlet",urlPatterns = "/userInfo.do")
public class UserInfoServlet extends HttpServlet {
    private static final long serialVersionUID = 1L;
    @Override
    protected void doPost(HttpServletRequest req, HttpServletResponse 
resp) throws ServletException, IOException {
        doGet(req,resp); }
    @Override
    protected void doGet(HttpServletRequest req, HttpServletResponse resp) 
throws ServletException, IOException {
        HttpSession session=req.getSession();
        User user1= (User) session.getAttribute("user");
        IUserService userService=new UserServiceImpl();
        User user=userService.getUserByUser(user1.getUsername());
        req.setAttribute("user",user);
```

```
            req.getRequestDispatcher("userinfo.jsp").forward(req,resp);
        }
    }
```

【代码说明】

当登录用户在 left.jsp 页单击个人中心链接时，就会交给 UserInfoServlet 进行处理，首先得到 session 用户对象 user，通过调用业务逻辑对象类 userService.getUserByUser(user1.getUsername())把查询结果封装到 user 对象中。

2. 修改保存用户信息

在 com.qzmall.servlet.user 包下新建 UserInfoSaveServlet 类。具体代码如下：

```
@WebServlet(name = "UserInfoSaveServlet",urlPatterns = "/userInfoSave.do")
public class UserInfoSaveServlet extends HttpServlet {
    private static final long serialVersionUID = 1L;
    @Override
    protected void doPost(HttpServletRequest req, HttpServletResponse resp) throws ServletException, IOException {
        HttpSession session = req.getSession();
        IUserService userService = new UserServiceImpl();
        User user1= (User) session.getAttribute("user");
        String username=user1.getUsername();
        String truename = req.getParameter("truename");
        String phone = req.getParameter("phone");
        String address = req.getParameter("address");
        String birthday = req.getParameter("birthday");
        String postcode = req.getParameter("postcode");
        int question = Integer.parseInt(req.getParameter("question"));
        String answer = req.getParameter("answer");
            User user = new User();
            user.setUsername(username);
            user.setAddress(address);
            user.setPhone(phone);
            user.setTruename(truename);
            user.setQuestion(question);
            user.setAnswer(answer);
            user.setPostcode(postcode);
            user.setBirthday(birthday);
            if (userService.saveUserInfo(user)) {
                String result = "修改信息成功！";
            req.setAttribute("resu", result);
            req.getRequestDispatcher("userInfo.do").forward(req,resp);
            } else {
        String result = "修改信息失败！";
            req.setAttribute("resu", result);
            req.getRequestDispatcher("userInfo.do").forward(req, resp);
            }
        }
    }
```

【代码说明】

当登录用户在 userinfo.jsp 页提交修改的个人信息后，就会交给 UserInfoSaveServlet 进行处理，它首先用 req 对象得到从页面提交的用户信息，并把它封装到 user 对象中，然后调用业务逻辑对象类 userService.saveUserInfo(user)保存修改结果。

(四)用户个人信息子模块视图层设计——个人中心页面设计

1. 左边通用页(left.jsp)页面设计

代码如下：

```jsp
<%@ page contentType="text/html;charset=UTF-8" language="java" %>
<html>
<head>
    <link type="text/css" rel="stylesheet" href="css/left.css">
</head>
<body>
<ul class="nav-side">
    <li class="nav-item acive"> <a href="userInfo.do" class="link">个人中心</a></li>
    <li class="nav-item"> <a href="ordersInfo.do?currentPage=1" class="link">我的订单</a></li>
    <li class="nav-item"> <a href="myPoint.do?currentPage=1" class="link">我的积分</a></li>
    <li class="nav-item"> <a href="modifypass.jsp" class="link">修改密码</a></li>
</ul>
</body>
</html>
```

2. 用户个人信息页(userinfo.jsp)设计

代码如下：

```jsp
<%@ taglib prefix="c" url="http://java.sun.com/jsp/jstl/core" %>
<%@ page contentType="text/html;charset=UTF-8" language="java" %>
<!DOCTYPE html>
<html>
<head>
    <meta charset="utf-8" />
    <meta http-equiv="X-UA-Compatible" content="ie=edge">
    <title>个人中心</title>
    <link type="text/css" rel="stylesheet" href="css/font-awesome.css">
    <link type="text/css" rel="stylesheet" href="css/font-awesome.min.css">
    <link type="text/css" rel="stylesheet" href="css/reset.css">
    <link type="text/css" rel="stylesheet" href="css/userinfo.css">
    <script src="js/jquery-1.11.2.min.js"></script>
    <script type="text/javascript" src="js/My97DatePicker/WdatePicker.js"></script>
    <script type="text/javascript" src="js/userinfo.js"></script>
```

```html
</head>
<body>
<jsp:include page="head.jsp"/>
<div class="crumb">
    <div class="w">
        <div class="crumb-con">
            <a href="index.do" class="link">MMAIL</a>
            <span>></span>
            <span class="link-text">个人中心</span>
        </div>
    </div>
</div>
<div class="page-wrap w">
    <jsp:include page="left.jsp"/>
    <div class="content with-nav">
        <div class="panel">
        <div class="panel-title">个人中心   <span id="span12">${resu}</span></div>
            <div class="panel-body">
        <form name="formuserinfo" method="post" action="userInfoSave.do" onsubmit="return checkForm();">
            <div class="form-line">
            <span class="label">用户名</span>
            <span class="text"><input type="text" id="username" name="username" value="${user.username}" onblur="checkUsername();" onfocus="checkHideUsername();" readonly="readonly"/></span>
     <span id="span1"></span>
     <span class="label">实  名：</span>
     <span class="text"><input type="text" id="truename" name="truename" value="${user.truename}"></span></div>
                <div class="form-line">
    <span class="label">手  机：</span>
    <span class="text"><input type="text" id="phone" name="phone" value="${user.phone}"></span>
     <span class="label">生  日：</span>
     <span class="text">
     <input class="user-content" id="birthday" name="birthday" value="${user.birthday}" onclick="WdatePicker()";>
          </span> </div>
      <div class="form-line">
      <span class="label">问  题：</span>
           <span class="text">
     <select name="question" id="question"  class="quest">
     <option value="0" <c:if test="${user.question==0}">
     selected="selected"</c:if>>请输入密码保护问题</option>
     <option value="1" <c:if test="${user.question==1}">
     selected="selected"</c:if>>你的中学名字</option>
     <option value="2" <c:if test="${user.question==2}">
     selected="selected"</c:if>>你的母亲名字</option>
     <option value="3" <c:if test="${user.question==3}">
```

```
                selected="selected"</c:if>>你喜欢的人物是</option>
                <option value="4" <c:if test="${user.question==4}">
                selected="selected"</c:if>>你父亲的生日</option>
                    </select>
                        </span>
        <span class="label">答  案:</span>
        <span class="text"><input type="text" id="answer" name="answer"
value="${user.answer}"></span>
                        </div>
        <div class="form-line">
            <span class="label">地  址:</span>
            <span class="text"><input type="text" id="address"
name="address" value="${user.address}"></span>
                 <span class="label">注册日期:</span>
            <span class="text"><input type="text" id="adddate"
name="adddate" value="${user.adddate}" readonly="true"></span>
                    </div>
            <div class="form-line">
                <span class="label">邮  编:</span>
                <span class="text"><input type="text" id="postcode"
name="postcode" value="${user.postcode}"></span>
                    </div>
                <div class="form-line">
                <input type="submit" value="修   改"
class="btn btn-submit"/>                                </div>
            </form>
                </div>
            </div>
        </div>
</div>
    <jsp:include page="foot.jsp"/>
</body>
</html>
```

用户个人信息页面运行效果如图6-4所示。

图6-4 用户个人信息页面

【代码说明】

(1) 用户个人信息页(userinfo.jsp)主要显示登录用户的基本信息,它接收UserInfoServlet传回页面的user对象,然后采用EL标签进行解析。

(2) 在显示密码提示问题时，用表单的选择框<select></select>进行显示，既要显示用户注册时输入的密码，还要方便用户选择修改个人信息。这里显示列表项时，我们用了 JSTL 标签条件选择标签 <option value="1" <c:if test="${user.question==1}">selected="selected"</c:if>你的中学名字</option>，可以方便地达到上述要求。

(3) <span id="span12">${resu}</span></div>这段代码主要用于数据回显，提示用户更新信息是否成功。

如上：把毕业中学更改为"秦皇岛市一中"，单击"修改"按钮，出现如图 6-5 所示的页面提示，表示修改信息成功。

图 6-5　提示修改信息成功

课堂技能训练：

【实训操作内容】用户个人信息维护功能实现。

【实训操作要求】

(1) 持久层设计。

(2) 业务逻辑层设计。

(3) 控制层设计。

(4) 视图层设计。

<center>课 后 训 练</center>

一、选择题

1. 如果需要在 JSP 页面使用 map 集合，那么需要导入哪个包，怎么书写？（　　）

　　A. <%@ page import = "java.map.* " %>

　　B. <%@ page import=java.util.* %>

　　C. <%@ import= "java.util.* " %>

　　D. <%@ page import= "java.util.* " %>

2. 使用 request 对象动态地获取工程名的正确写法是（　　）。

　　A. request.getServletContext()　　　　B. request.getServerName()

C. request.getContextPath()　　　　D. request.getServletContext().getContextPath()

3. 编写 Servlet 的 service 方法时，需要抛出的异常是(　　)。

　　A. ServletException，RemoteException

　　B. ServletException，IOException

　　C. HttpServletException，RemoteException

　　D. HttpServletException，IOException

4. 在 JSP 表单中包含如下语句：<input name:"password">，表单提交后，想要在 Servlet 中获取到，应该使用 HttpServletRequest 对象的(　　)方法。

　　A. getParameter("password");　　　B. getParameter();

　　C. getAttribute ();　　　　　　　　D. getAttribute("password");

5. 关于转发，以下说法正确的是(　　)。

　　A. 转发时，浏览器中的地址栏 url 会发生变化

　　B. 转发调用的是 HttpServletRequest 对象中的方法

　　C. 转发调用的是 HttpServletResponse 对象中的方法

　　D. 转发时浏览器只请求一次服务器

6. 使用重定向时，需要将数据带到跳转页面，可以使用(　　)对象。

　　A. request　　　B. session　　　C. page　　　D. application

7. request 对象可以使用(　　)方法获取表单中某输入框提交的信息，并且返回值为 String 类型。

　　A. getParameter(String s)　　　　B. getValue(String s)

　　C. getParameterNames(String s)　　D. getParameterValues(String s)

8. 在 a.jsp 页面中有如下代码：pageContext.setAttribute("name",Smith)，同时跳转到 b.jsp，那么如何在 b.jsp 页面中获取到 name 的值？(　　)

　　A. pageContext.getAttribute("name")

　　B. pageContext.getParameter("name")

　　C. 其余说法均不正确

　　D. request.getAttribute("name")

9. 关于 session 域的说法错误的是(　　)。

　　A. 可以调用 HttpSession 的 invalidate 方法，立即销毁 session 域

　　B. 当 Web 应用被移除出 Web 容器时，Web 容器对应的 session 跟着销毁

　　C. session 域的作用范围为整个会话

　　D. session 域中的数据只能存在 30 分钟，这个时间不能修改

10. config 内置对象获取配置文件的初始化参数的正确写法是(　　)。

　　A. config.getAtrribute()　　　　　B. config.getParameter()

　　C. config.getInit()　　　　　　　　D. config.getInitParameter()

11. 在 jsp 页面中如何获取响应的字符编码集？（ ）

    A. response.setCharacterEncoding()　　　　B. request. getCharacterEncoding()

    C. response.getCharacterEncoding()　　　　D. request. setCharacterEncoding()

二、实际操作题

进行用户模块的持久层、业务逻辑层、控制层、视图层方法设计，完成用户个人积分的查看、个人密码的修改功能。

第六章 习题答案.docx

第六章 用户模块设计.pptx

第七章

购物车与订单模块设计

知识能力目标

1. 掌握 LinkedHashMap 集合类的使用方法。
2. 掌握购物车的主要功能和实现流程。
3. 掌握订单的主要构成要素。
4. 学会将商品添加到购物车及在购物车中删除商品、修改商品数量与清空购物车等方法。
5. 学会生成订单、查询订单及修改订单状态的设计方法。

问题提示

购物车功能指的是应用于网店的在线购买功能，它类似于超市购物时使用的推车或篮子，可以暂时把挑选的商品放入购物车、删除或更改购买数量，并对多件商品进行一次性结款，是网上商店里的一种快捷购物工具，使用购物车购物是需要进行登录的，只有这样才能进行结账，生成订单，并把货物及时送达消费者手中。

问题：如果用户没有进行登录，并访问了购物页，这时候系统该如何处理？

第一节 购物车模块设计

一、购物车的基本流程

（一）添加商品到购物车的基本流程

（1）接收前端传来的商品信息，如果购物车中不包含要添加的商品信息，直接将商品信息添加到购物车中；如果购物车中包含了要添加的商品信息，则更改商品数量即可。

（2）重新计算总计。

（二）从购物车中移除商品的流程

将商品信息从购物车中移除，重新计算总计。

（三）清空购物车的基本流程

将所有的商品信息从购物车中移除，重新计算总计。

二、购物车模块设计

可以随便打开一个购物页面，不难发现，购物车包括的信息主要是商品信息(包括商品名称、图片、价格)、商品数量、小计。另外在购物时，需要对其购物项进行增加、删除、修改，所以首先需要封装一个购物项类，然后通过调用这个类，完成购物车的各项功能。

(一)购物车类设计

1. 在 com.qzmall.cart 包下新建类 CartItem

代码如下：

```java
public class CartItem {
    private Shop shop;//购物项中商品信息
    private int quantity;//商品数量
    private float subTotal;//小计
    public Shop getShop() {
        return shop;   }
    public void setShop(Shop shop) {
        this.shop = shop;   }
    public int getQuantity() {
        return quantity;   }
    public void setQuantity(int quantity) {
        this.quantity = quantity;   }
    public float getSubTotal() {
        return quantity*shop.getPrice();  }
    public void setSubTotal(float subTotal) {
        this.subTotal = subTotal;}
}
```

2. 在 com.qzmall.cart 包下新建类 Cart

代码如下：

```java
public class Cart {
    //购物项集合：Map 的 Key 就是商品 ID，value 就是购物项
    private Map<Integer,CartItem> map=new LinkedHashMap<Integer,CartItem>();
    //购物总计
    private float total;
    public void setTotal(float total) {
        this.total = total;    }
    public float getTotal() {
        return total;   }
    //cart 对象中有一个 CartItems 的属性
    public Collection<CartItem> getCartItems(){
        return map.values();       }
    public boolean addCart(CartItem cartItem) {
        Integer pid = cartItem.getShop().getPid();
        //如果存在该商品
        IShopService shopService = new ShopServiceImpl();
        Shop shop = new Shop();
        shop = shopService.getShopByid(pid);
        if (map.containsKey(pid)) {
            CartItem _cartiItem = map.get(pid);
            if (_cartiItem.getQuantity() + cartItem.getQuantity() <= shop.getStock()) {
```

```java
                _cartiItem.setQuantity(_cartiItem.getQuantity() + cartItem.getQuantity());
                total += cartItem.getSubTotal();
                return true;
            } else {
                return false;
            }
        }
        //如果不存在该商品
        else {
            if (cartItem.getQuantity() <= shop.getStock()) {
                map.put(pid, cartItem);
                total += cartItem.getSubTotal();
                return true;
            } else {
                return false;
            }
        }
    }
    //从购物车清除购物项
    public boolean remocart(Integer pid){
        //将购物项移出购物车
        CartItem cartItem=map.remove(pid);
        //减掉已移除购物项的小计
        total-=cartItem.getSubTotal();
        return true;     }
    public boolean updatecart(int pid,int quantity){
        IShopService shopService = new ShopServiceImpl();
        Shop shop = new Shop();
        shop = shopService.getShopByid(pid);
        //如果存在该商品
        if (map.containsKey(pid)){
            if(quantity<=shop.getStock()){
            CartItem _cartiItem=map.get(pid);
            Float ss=_cartiItem.getSubTotal();
            _cartiItem.setQuantity(quantity);
            total=total+_cartiItem.getSubTotal()-ss;
            return true;}
            else{
                return false;
              }
        }
        //如果不存在该商品
        else{
          return false;
        }
    }
    //清空购物车
    public boolean clearCart(){
        map.clear();
```

```
        total=0;
        return true;
    }
}
```

【代码说明】

首先，用 map 集合对购物项 CartItem 进行了封装，这样，增加、删除购物项比较方便。

其次，页面只需要一个购物项的集合，所以没有必要对 map 集合的所有项都生成 set、get 方法，那样遍历比较烦琐。只需要把 map 的 value 转成一个单列的集合就可以了，所以用方法 public Collection<CartItem> getCartItems(){return map.values(); }就可以把购物项全部商品遍历出来了。

最后，要求购物数量必须小于或等于该商品库存数量，所以购物数量小于或等于该商品库存数量时，返回真，否则返回假。

> **小提示：如何测试类方法**
>
> 写好了一个类方法之后，这个方法到底有没有错误，一般不能等到网站运行的时候才进行检验，在 IDEA 中，通常就在 com.qzmall.test.DBUtilTest 中进行测试。
>
> 下面测试一下上述购物车类中的 addCart()方法：
> ```
> public class DBUtilTest {
>     public static void main(String[] args) {
>         CartItem cartItem=new CartItem();
>         cartItem.setQuantity(2);
>         Shop shop=new Shop();
>         shop.setPid(17);
>         shop.setShopname("飞利浦电视");
>         shop.setPrice(29999f);
>         cartItem.setShop(shop);
>         Cart cart=new Cart();
>         System.out.println(cart.addCart(cartItem));}
> }
> ```
> shop 表里设定飞利浦电视库存数量为 5，购物数量为 2 时，返回 true，购物数量为 6 时，返回假，这说明 addCart 运行成功。

(二)购物车模块控制层设计

1. 在 com.qzmall.servlet.orders 包下添加 servlet:CartServlet

代码如下：

```
@WebServlet(name = "CartServlet",urlPatterns = "/cartServlet.do")
public class CartServlet extends HttpServlet {
    private int pid;
    private int quantity;
    private String action;  //表示购物车的动作, add, show, delete
```

```java
    //商品业务逻辑类的对象
    IShopService shopService = new ShopServiceImpl();
    public void doPost(HttpServletRequest request, HttpServletResponse response) throws ServletException, IOException {
        doGet(request, response);       }
    public void doGet(HttpServletRequest request, HttpServletResponse response) throws ServletException, IOException {
        //System.out.println("进入CartServlet");
        PrintWriter out = response.getWriter();
        String action=request.getParameter("action");
        if (action != null) {
            if (action.equals("add"))   //如果是添加商品
            {
                if (addToCart(request, response))
{ request.getRequestDispatcher("/cart.jsp").forward(request, response);
                } else {
                    String msg = "添加购物车失败！";
                    request.setAttribute("msg", msg);
request.getRequestDispatcher("/result.jsp").forward(request, response);
                }
            }
            }
            if (action.equals("delete"))   //如果是执行删除
            {
                if (deleteFromCart(request, response))
{ request.getRequestDispatcher("/cart.jsp").forward(request, response);
                } else {
                String result = "删除购物项失败！";
                 request.setAttribute("msg", result);
request.getRequestDispatcher("/result.jsp").forward(request, response);
                }
            }
            if (action.equals("clear"))   //如果是执行清空
            {
                Cart cart = (Cart) request.getSession().getAttribute("cart");
                if (cart.clearCart())
{ request.getRequestDispatcher("/cart.jsp").forward(request, response);
                } else {
                    String result = "清空购物车失败！";
                    request.setAttribute("msg", result);
request.getRequestDispatcher("/reault.jsp").forward(request, response);
                }
            }
        if (action.equals("update"))   //如果是执行删除购物车中的商品
            {
            if (request.getSession().getAttribute("cart") == null) {
 Cart cart = new Cart();
request.getSession().setAttribute("cart", cart);
            }
```

```java
        Cart cart = (Cart) request.getSession().getAttribute("cart");
        int pid = Integer.parseInt(request.getParameter("pid"));
        int quantity = Integer.parseInt(request.getParameter("quantity"));
            Shop shop = shopService.getShopByid(pid);
            if (quantity <=shop.getStock()) {
                CartItem cartItem = new CartItem();
                //设置数量
                cartItem.setQuantity(quantity);
                //设置商品
                cartItem.setShop(shop);
                if (cart.updatecart(pid,quantity)) {
                    StringBuilder sb = new StringBuilder("{");
sb.append("\"quantity\"").append(":").append(cartItem.getQuantity());
                    sb.append(",");
sb.append("\"subtotal\"").append(":").append(cartItem.getSubTotal());
                    sb.append("}");
                    response.getWriter().print(sb);
                }
            } else {
            String result = "购物车商品数量更新失败！";
                request.setAttribute("msg", result);
request.getRequestDispatcher("/reault.jsp").forward(request, response);
            }
        }
    }
}
    //添加商品到购物车的方法
private boolean addToCart(HttpServletRequest request,
HttpServletResponse response) {
        //是否是第一次给购物车添加商品，需要给session中创建一个新的购物车对象
     if (request.getSession().getAttribute("cart") == null) {
        Cart cart = new Cart();
        request.getSession().setAttribute("cart", cart);
     }
     Cart cart = (Cart) request.getSession().getAttribute("cart");
     String pid = request.getParameter("pid");
     String quantity = request.getParameter("quantity");
     Shop shop = shopService.getShopByid(Integer.parseInt(pid));
     CartItem cartItem = new CartItem();
     //设置数量
    cartItem.setQuantity(Integer.parseInt(quantity));
     //设置商品
    cartItem.setShop(shop);
     if (cart.addCart(cartItem)) {
        return true;
     } else {
        return false;
     }
    }
    //从购物车中删除商品
```

```java
    private boolean deleteFromCart(HttpServletRequest request,
HttpServletResponse response)
    {
       int pid =Integer.parseInt( request.getParameter("pid"));
       Cart cart = (Cart)request.getSession().getAttribute("cart");
       if(cart.remocart(pid))
       {
           return true;
       }
       else
       {
           return false;
       }
    }
}
```

【代码说明】

购物车操作包括四种动作，分别是添加购物项(add)、删除购物项(delete)、更新商品数量(update)、清空购物车(clear)，通过在页面传递一个 action 参数来分别表示四种动作，这样就可以在一个 CartServlet 类中完成上述四种操作。

2. 在 web.xml 中配置 CartServlet

代码如下：

```xml
<!-- 限制非登录用户进入 cart.jsp-->
<filter-mapping>
    <filter-name>validatelogn</filter-name>
    <url-pattern>/cart.jsp</url-pattern>
</filter-mapping>
<!-- 限制非登录用户使用 cartServlet.do-->
<filter-mapping>
    <filter-name>validatelogn</filter-name>
    <url-pattern>/cartServlet.do</url-pattern>
</filter-mapping>
```

(三)视图层设计

1. 在 cart.jsp 页面显示购物信息

代码如下：

```jsp
<%@ taglib prefix="c" uri="http://java.sun.com/jsp/jstl/core" %>
<%@ page contentType="text/html;charset=UTF-8" language="java" %>
<!DOCTYPE html>
<html>
<head>
    <meta charset="UTF-8">
    <meta http-equiv="X-UA-Compatible" content="ie=edge">
    <title>购物车</title>
    <link type="text/css" rel="stylesheet" href="css/font-awesome.css">
    <link type="text/css" rel="stylesheet" href="css/reset.css">
```

```html
        <link type="text/css" rel="stylesheet" href="css/cart.css">
        <script src="https://libs.baidu.com/jquery/2.1.4/jquery.min.js"></script>
        <script src="js/cart.js"></script>
</head>
<body>
<jsp:include page="head.jsp"/>
<div class="page-wrap w">
    <table id="cartTable">
        <thead>
        <tr>
            <th>商品</th> <th>单价</th>
            <th>数量</th> <th>小计</th> <th>操作</th>
        </tr>
        </thead>
        <tbody>
        <c:forEach items="${sessionScope.cart.cartItems}" var="cartItem">
        <tr>
            <td class="goods"><img src="<c:url value='img/pic/$ {cartItem.shop.image1}'/>"><span>${cartItem.shop.shopname}</span></td>
            <td class="price">${cartItem.shop.price}</td>
            <td class="count">
                <input id="${cartItem.shop.pid}Stock" value="${cartItem.shop.stock}" type="hidden"/>
                <a class="jian" id="${cartItem.shop.pid}jian">-</a><input id="${cartItem.shop.pid}Quantity" type="text" value="${cartItem.quantity}" readOnly="true"/><a class="jia" id="${cartItem.shop.pid}jia">+</a>
            </td>
            <td class="subtotal" id="${cartItem.shop.pid}"><span id="${cartItem.shop.pid}Subtotal">${cartItem.subTotal}</span></td>
            <td class="operation"><span class="delete"><a href="cartServlet.do?pid=${cartItem.shop.pid}&action=delete">删除</a></span></td>
        </tr>
         </c:forEach>
        </tbody>
    </table>
    <div class="foot" id="foot">
        <a class="btn " id="deleteAll" href="cartServlet.do?action=clear">清空购物车</a>
        <input class="btn" type="button" value="返回继续购物" onClick="javascript:window.close()">
        <div class="fr "><a class="btn jiesuan" id="jiesuan" href="orderServlet.do">生成订单</a></div>
        <div class="fr total">合计：¥<span id="total">${sessionScope.cart.total}</span></div>
    </div>
</div>
<jsp:include page="foot.jsp"/>
</body>
```

```
</html>
```

【代码说明】

页面利用 JSTL 标签对 CartServlet 传来的 session 集合对象 cart 进行解析。Cart 对象包括两个部分：购物项(cartItem)和购物总额(total)。购物项内容包括商品实体对象 shop 和商品数量(quantity)、商品小计(subTotal)，弄清了这几个层次关系，对 cart 对象进行遍历就一清二楚了。

2. cart.jsp 代码设计

代码如下：

```
$(function () {
    function showTotal() {
        var total = 0;
        /*
        1. 获取所有的被勾选的条目复选框！对它们进行循环遍历
        */
        $(".subtotal").each(function(){
            var id=$(this).attr("id");
            var text = $("#" + id + "Subtotal").text();
            //2. 累加计算
            total += Number(text);
        })
        // 3. 把总计显示在总计元素上
        $("#total").text(total); }
    showTotal();
    $(".jian").click(function(){
        var dd= $(this).attr("id").indexOf("jian")
        var pid=$(this).attr("id").substring(0,dd);
        var quantity=$("#"+pid+"Quantity").val();
        alert(quantity);
        if(quantity==1){
            alert("采购数量不能低于1")
            return false;
        }
        else{
            sendUpdateQuantity(pid,Number(quantity)-1);
        }
});
    $(".jia").click(function(){
        var dd= $(this).attr("id").indexOf("jia")
        var pid=$(this).attr("id").substring(0,dd);
        var quantity=$("#"+pid+"Quantity").val();
        var stock=$("#"+pid+"Stock").val()
        if (Number(quantity)>Number(stock)-1){
            alert("采购数量大于库存！")
            return false;
        }
```

```
            sendUpdateQuantity(pid,Number(quantity)+1)
});
    function sendUpdateQuantity(pid, quantity) {
        $.ajax({
            url : "cartServlet.do",
            type : "POST",
            data : {
                "action" :"update",
                "pid" : pid,
                "quantity":quantity,
                "time" : new Date().getTime()
            },
            dataType : "json",
            success:function(result) {
                //1. 修改数量
                $("#" + pid + "Quantity").val(result.quantity);
                //2. 修改小计
                $("#" + pid + "Subtotal").text(result.subtotal);
                //3. 重新计算总计
                showTotal();
            }
        });
    }
})
```

【代码说明】

在"数量"这一栏，在数量字段旁边加了两个按钮，目的是通过单击这两个按钮进行数量修改。另外，还要注意修改商品数量后，需要重新计算小计和总计，页面不需要刷新，所以这样采用 ajax()方法，通过 sendUpdateQuantity()函数调用 cartServlet.do，通过往其传递了三个参数：pid、action、quantity，这样就可以把修改后的小计和总计无刷新地传回界面。运行效果如图 7-1 所示。

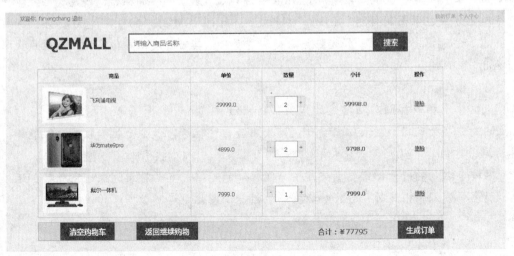

图 7-1　购物车页面(cart.jsp)

课堂技能训练：

【实训操作内容】 完成购物车类设计。

【实训操作要求】 要求购物类具有以下功能。

(1) 增加商品到购物车功能。

(2) 删除商品功能。

(3) 清空购物车功能。

(4) 异步请求更新购物车功能。

【知识拓展】

实现购物车的方法有很多，本例中我们主要是通过 session 对象实现了购物车的各项功能，也可以采用其他方式，这些方式各有优劣，介绍如下。

(1) cookie。

cookie 存储在客户端，且占用很少的资源，浏览器允许存放 300 个 cookie，每个 cookie 的大小为 4KB，足以满足购物车的要求，同时也减轻了服务器的负荷。

cookie 为浏览器所内置，使用方便。即使用户不小心关闭了浏览器窗口，只要在 cookie 定义的有效期内，购物车中的信息也不会丢失。

基于 cookie 的购物车要求用户浏览器必须支持并设置为启用 cookie，否则购物车则失效；存在着关于 cookie 侵犯访问者隐私权的争论，因此有些用户会禁止本机的 cookie 功能。

(2) session。

session 是实现购物车的另一种方法。它与 cookie 最重大的区别是，session 将用户在会话期间的私有信息存储在服务器端，提高了安全性。在服务器生成 session 后，客户端会生成一个 sessionid 识别号保存在客户端，以保持和服务器的同步。这个 sessionid 是只读的，如果客户端禁止 cookie 功能，session 会通过在 URL 中附加参数，或隐含在表单中提交等其他方式在页面间传送。因此，利用 session 实施对用户的管理更为安全、有效。session 会占用服务器资源，加大服务器端的负载，尤其当并发用户很多时，会生成大量的 session，影响服务器的性能。

(3) 结合数据库的方式。

数据库承担着存储购物信息的作用，session 或 cookie 则用来跟踪用户。数据库与 cookie 分别负责记录数据和维持会话，能发挥各自的优势，使安全性和服务器性能都得到提高。

每一个购物的行为，都要直接建立与数据库的连接，直至对表的操作完成后，连接才释放。当并发用户很多时，会影响数据库的性能，因此，这对数据库的性能提出了更高的要求。

第二节 生成订单子模块设计

一、生成订单模块的基本流程

(一)生成主订单信息的流程

(1) 得到用户提交的信息(收货人、收货地址、邮编、手机号)和订单总额。
(2) 生成订单编号(保证订单编号不会重复)。
(3) 设定订单初始状态为1。
(4) 生成订单主表。
(5) 扣除会员积分。

(二)生成订单明细的流程

(1) 得到用户购买的商品信息(商品编号、商品名称、商品数量)。
(2) 把每一商品信息和订单编号一并加入订单明细表。

二、生成订单子模块功能实现

(一)生成订单子模块持久层设计

1. 接口类设计

在 com.qzmall.dao 包下新建接口类 IOrdersDao：

```java
public interface IOrdersDao {
    //生成新的订单
    public boolean addOrders(Orders orders);
    //生成订单明细
    public boolean addOrdersItem(OrdersItem ordersItem);
}
```

2. 接口实现类设计

在 com.qzmall.dao.impl 包下新建接口实现类 OrderDaoImpl：

```java
public class OrderDaoImpl implements IOrdersDao {
/*
```

(1) 生成新的订单。

该方法用来实现向订单主表 orders 添加一条记录，以订单对象为参数，返回结果为布尔类型，表示生成新的订单是否成功。具体代码如下：

```java
*/
@Override
public boolean addOrders(Orders orders) {
```

```
    String strSql = "insert into orders
(orderId,username,address,postcode,truename,addtime,phone,sum,state)
values (?,?,?,?,?,?,?,?,?) ";
 return DBUtil.executeUpdate(strSql, orders.getOrderId(),
orders.getUsername(), orders.getAddress(), orders.getPostcode(),
orders.getTruename(), orders.getAddtime(), orders.getPhone(),
orders.getSum(), orders.getState()) > 0;}
 /*
```

(2) 生成新的订单明细。

该方法用来实现向订单明细表 ordersitem 添加一条记录,以订单明细对象(ordersItem)为参数,返回结果为布尔类型,表示生成新的订单是否成功。具体代码如下:

```
 */
    @Override
    public boolean addOrdersItem(OrdersItem ordersItem) {
    String strSql = "insert into ordersitem
(orderId,pid,shopname,price,shopnum) values (?,?,?,?,?) ";
    return DBUtil.executeUpdate(strSql, ordersItem.getOrderId(),
ordersItem.getPid(), ordersItem.getShopname(), ordersItem.getPrice(),
ordersItem.getShopnum()) > 0;
    }
}
```

(二)生成订单子模块逻辑层设计

1. 接口类设计

接口类中的方法与持久层一样,这里不再赘述。

2. 接口实现类设计

代码如下:

```
 */
public class OrdersServiceImpl implements IOrdersService {
    @Override
    public boolean addOrders(Orders orders) {
        IOrdersDao ordersDao=new OrderDaoImpl();
        return ordersDao.addOrders(orders);
    }
    @Override
    public boolean addOrdersItem(OrdersItem ordersItem) {
        IOrdersDao ordersDao=new OrderDaoImpl();
        return ordersDao.addOrdersItem(ordersItem);
    }
}
```

(三)生成订单子模块控制层设计

1. 跳转到生成订单页面(order.jsp)

当登录用户从 cart.jsp 页发出生成商品的请求时,系统就会交给 OrderServlet 来处理,该类通过调用业务逻辑类 userService.getUserByUser 方法得到用户对象 user，然后把它传给订单页面。具体代码如下：

```
@WebServlet(name = "OrderServlet",urlPatterns = "/orderServlet.do")
public class OrderServlet extends HttpServlet {
    protected void doGet(HttpServletRequest req, HttpServletResponse resp)
throws ServletException, IOException { HttpSession
session=req.getSession();
    User user1= (User) session.getAttribute("user");
IUserService userService=new UserServiceImpl();
User user=userService.getUserByUser(user1.getUsername());
    req.setAttribute("user",user);
req.getRequestDispatcher("order.jsp").forward(req,resp);
    }
}
```

2. 订单结算功能设计

当登录用户从 order.jsp 页发出订单结算的请求时，系统就会交给 CheckOutServlet 来处理。具体代码如下：

```
@WebServlet(name = "CheckOutServlet",urlPatterns = "/checkOut.do")
public class CheckOutServlet extends HttpServlet {
protected void doPost(HttpServletRequest req, HttpServletResponse resp)
throws ServletException, IOException {    HttpSession session =
req.getSession();
        int jf=0;
        IUserService userService=new UserServiceImpl();
        IOrdersService ordersService=new OrdersServiceImpl();
        User user= (User) session.getAttribute("user");
        Cart cart=(Cart)session.getAttribute("cart");
        Orders orders=new Orders();
        OrdersItem ordersItem=new OrdersItem();
        float sum=cart.getTotal();
        long adddate =StringHelper.getCurrentTimeStamp();
        Random rand =new Random();
        int i=rand.nextInt(1000);
        adddate=adddate*1000+i;
        String orderId=String.valueOf(adddate);
        String address=req.getParameter("address");
        String phone=req.getParameter("phone");
        String postcode=req.getParameter("postcode");
        String truename=req.getParameter("truename");
        String username=user.getUsername();
        int point=Integer.parseInt(req.getParameter("point"));
        String Isjifen=req.getParameter("Isjifen");
        if (Isjifen!=null && !"".equals(Isjifen)) {
            int jifen1 = Integer.parseInt(req.getParameter("jf1"));
            System.out.println("jifen1="+jifen1);
```

```java
            if (jifen1 <= point) {
                jf = jifen1 / 1000;
                sum = sum - jf;
                req.setAttribute("sum",sum);
                String sql = "兑换积分";
                Point point1=new Point();
                point1.setUsername(username);
                point1.setPointtime(StringHelper.getCurrentFormatDate());
                point1.setOrderid(orderId);
                point1.setMypoint(-jf*1000);
                point1.setDescription(sql);
                userService.addPoint(point1);
            }
            else{
                return;
            }
        }
        orders.setOrderId(orderId);
        orders.setUsername(username);
        orders.setAddress(address);
        orders.setPhone(phone);
        orders.setAddtime(StringHelper.getCurrentFormatDate());
        orders.setTruename(truename);
        orders.setPostcode(postcode);
        orders.setSum(sum);
        orders.setState(1);//1.未付款 2.已付款，未发货 3.已发货 4.交易完成
        if (ordersService.addOrders(orders)){
            for (CartItem cartItem:cart.getCartItems()){
            ordersItem.setShopnum(cartItem.getQuantity());
                ordersItem.setPrice(cartItem.getShop().getPrice());
                ordersItem.setShopname(cartItem.getShop().getShopname());
                ordersItem.setPid(cartItem.getShop().getPid());
            ordersItem.setOrderId(orderId);
                System.out.println(ordersService.addOrdersItem(ordersItem));
             }
            req.setAttribute("orderId",orderId);
            String msg="";
            if (Isjifen!=null)
            { msg="你的订单编号为：'"+orderId+"', 金额为"+cart.getTotal()+"元, 扣除会员积分"+jf*1000+", 应付"+sum+"元";
            }
            else{
             msg="你的订单编号为：'"+orderId+"', 金额为"+cart.getTotal()+"元, 应付"+sum+"元";
            }
            session.removeAttribute("cart");
            req.setAttribute("msg",msg);
        req.getRequestDispatcher("checkout.jsp").forward(req,resp);
```

```
        }
        else{
            String msg="发生异常,订单没有完成";
            req.setAttribute("msg",msg);
            req.getRequestDispatcher("result.jsp").forward(req,resp);
        }
    }
    protected void doGet(HttpServletRequest req, HttpServletResponse resp)
    throws ServletException, IOException { doPost(req,resp);}
}
```

【代码说明】

订单结算其实就是往订单主表和订单明细表中添加记录。

(1) 我们要从 orders.jsp 页面通过 req 对象得到如收货地址(address)、真实姓名(truename)、手机号码(phone)、邮编(postcode)、总计(sum)、用户积分(point)这些信息。

(2) 保证订单主表(orders)的订单编号(orderId)不会重复。

由于订单在订单主表中是不能重复的,所以采用生成订单的实时时间和随机数生成订单编号,调用 StringHelper.getCurrentTimeStamp()方法生成时间的长整型,又通过 int i=rand.nextInt(1000)得到一个三位数的随机数,然后把这两个数合并到一块儿,就可以保证订单编号不会重复。

(3) 订单状态是初始状态,设为 1;订单生成时间是系统默认时间。

(4) 如果会员使用会员积分,系统规定 1000 个积分折合 1 元钱,使用会员积分必须是 1000 的整数位,结算订单总额时要扣除会员积分相抵的金额,同时要在积分表中添加积分使用信息。

(5) 把订单编号、收货地址、真实姓名、订单金额、订单状态等信息封装到订单主表实体对象 order 中,把会员积分信息封装到 point1 对象中,利用订单业务逻辑类生成主订单信息,利用积分业务逻辑类生成用户积分记录,在生成订单主表信息的同时,还要把这一订单编号下的商品、数量、价格等信息逐一封装到订单明细表实体对象中,然后调用订单业务逻辑类生成订单明细信息。

> 小提示:Java Servlet API 中 forward() 与 redirect()的区别
>
> 前者仅是容器中控制权的转向,在客户端浏览器地址栏中不会显示出转向后的地址;后者则是完全的跳转,浏览器将会得到跳转的地址,并重新发送链接请求。这样,从浏览器的地址栏中可以看到跳转后的链接地址。因此,前者更加高效,在前者可以满足需要时,尽量使用 forward()方法,并且,这样也有助于隐藏实际的链接。在有些情况下,比如,需要跳转到一个其他服务器上的资源,则必须使用 sendRedirect()方法。

(四)生成订单子模块视图层设计

1. 生成订单(order.jsp)页面设计

代码如下:

```jsp
<%@ taglib prefix="c" uri="http://java.sun.com/jsp/jstl/core" %>
<%@ page contentType="text/html;charset=UTF-8" language="java" %>
<!DOCTYPE html>
<html>
<head>
    <meta charset="UTF-8">
    <meta http-equiv="X-UA-Compatible" content="ie=edge">
    <title>我的订单</title>
    <link type="text/css" rel="stylesheet" href="css/font-awesome.css">
    <link type="text/css" rel="stylesheet" href="css/reset.css">
    <link type="text/css" rel="stylesheet" href="css/order.css">
    <script src="https://libs.baidu.com/jquery/2.1.4/jquery.min.js"></script>
    <script type="text/javascript" src="js/order.js"></script>
</head>
<body>
<jsp:include page="head.jsp"/>
<div class="page-wrap w">
    <form id="form1" name="form1" action="checkOut.do" onsubmit="return sendreg();">
        <div class="panel">
            <h1 class="panel-title">收货地址</h1>
            <div class="panel-body">
                <div class="form-line">
                <span class="label">实　名：</span>
                <span class="text"><input type="text" id="truename" name="truename" value="${user.truename}"></span>
                <span class="label">手　机：</span>
                <span class="text"><input type="text" id="phone" name="phone" value="${user.phone}"></span>
                </div>
                <div class="form-line">
                <span class="label">地　址：</span>
                <span class="text"><input type="text" id="address" name="address" value="${user.address}" class="address"></span>
                </div>
                <div class="form-line">
                <span class="label">积　分：</span>
                <span class="text"><input type="text" id="point" name="point" value="${user.point}" readOnly="true"></span>
                <span class="label">邮　编：</span>
                <span class="text"><input type="text" id="postcode" name="postcode" value="${user.postcode}"></span>
                </div>
            </div>
            <h1 class="panel-title">商品清单</h1>
```

```html
                <div class="panel-body">
                  <table class="product-table">
                    <tr>
                        <th class="cell-img">商品图片</th>
                        <th class="cell-info">商品名称</th>
                        <th class="cell-price">价格</th>
                        <th class="cell-count">数量</th>
                        <th class="cell-total">小计</th>
                    </tr>
            <c:forEach items="${sessionScope.cart.cartItems}" var="cartItem">
              <tr>
                  <td class="cell-img"><a href=""><img src="<c:url value='img/pic/${cartItem.shop.image1}'/>" class="p-img"/></a></td>
                  <td class="cell-info"><a href="">${cartItem.shop.shopname}</a></td>
                  <td class="cell-price">¥${cartItem.shop.price}</td>
                  <td class="cell-count">${cartItem.quantity}</td>
                  <td class="cell-total">${cartItem.subTotal}</td>
              </tr>
            </c:forEach>
         </table>
         <dl class="jifen">
           <dt><input name="Isjifen" type="checkbox" id="Isjifen" onClick="javascript:ChkIsjifen();" value="1" />兑换积分</dt>
           <dd id="Isjifen1"  style="display:none;"><input name="jf1" type="text" id="jf1"  onblur="javascript:return test();"/>填写不能超过本人积分,且必须是1000的倍数</dd>
         </dl>
         <div class="submit-con">
         <span>订单总价: </span>
         <span class="submit-total">${sessionScope.cart.total} </span>
         <span class="btn order-submit"><input type="submit" value="提交订单"/></span>
         </div>
         </div>
      </div>
    </form>
</div>
<jsp:include page="foot.jsp"/>
</html>
```

【代码解析】
该页面是根据购物车里的商品生成订单内容,这个页面包括以下三个方面的内容。
(1) 订单基本信息。利用 EL 标签解析 OrderServlet 传回的 user 对象生成这部分内容。
(2) 订单明细信息。利用 JSTL 标签对 session 对象 cart 遍历生成其内容。
(3) 兑换积分信息。这个可以由用户自己来进行选择,积分填写时不能超过本人积分，且必须是 1000 的倍数。运行效果如图 7-2 所示。

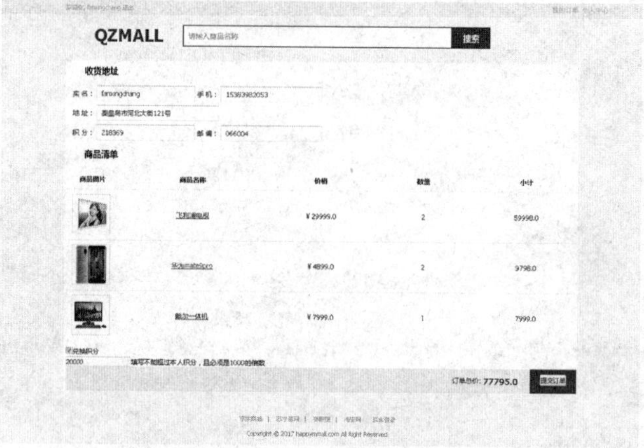

图 7-2 生成订单(order.jsp)页面

2. 提交订单(checkout.jsp)页面设计

代码如下：

```jsp
<%@ page contentType="text/html;charset=UTF-8" language="java" %>
<!DOCTYPE html>
<html>
<head>
    <meta charset="UTF-8">
    <meta http-equiv="X-UA-Compatible" content="ie=edge">
    <title>秦职电商平台</title>
    <link type="text/css" rel="stylesheet" href="css/font-awesome.css">
    <link type="text/css" rel="stylesheet" href="css/reset.css">
    <link type="text/css" rel="stylesheet" href="css/result.css">
</head>
<body>
<jsp:include page="head.jsp"/>
<div class="page-wrap w">
    <div class="result-con">
        <h1 class="result-title">${msg}</h1>
        <div class="result-content">
          <a href="index.do" class="link">回到首页</a>
        </div>
    </div>
</div>
<jsp:include page="foot.jsp"/>
</body>
</html>
```

【代码说明】

页面主要显示订单结算后的信息，由 EL 标签解析 CheckOutServlet 传回的 msg 对象生成其内容，主要包括订单编号、订单金额、会员应付金额。运行效果如图 7-3 所示。

图 7-3　提交订单(checkout.jsp)页面

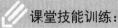课堂技能训练：

【实训操作内容】完成订单子模块的结算功能。

【实训操作步骤】

(1) 持久层设计。
(2) 业务逻辑层设计。
(3) 控制层设计。
(4) 视图层设计。

第三节　我的订单子模块设计

一、我的订单子模块基本流程

订单浏览基本流程：用户登录浏览我的订单。

订单查询基本流程：用户登录；输入订单编号；查询订单。

订单状态修改基本流程：用户登录；浏览我的订单、订单明细；修改订单状态。

取消订单基本流程：用户登录；浏览我的订单；选中某一订单；取消订单。

二、我的订单子模块主要功能实现

(一)我的订单子模块持久层设计

1. 接口类设计

在 com.qzmall.dao.IOrdersDao 接口类下新增以下方法：

```java
public interface IOrdersDao {
    // 分页查询用户订单信息
    public List<Orders> getOrdersPageByUser(String username, int currentPage, int pageSize);
    public int getOrdersTotalByUser(String username);
    // 根据订单编号查询订单主表详细信息
    public Orders getOrdersByOrderId(String orderId);
    // 根据订单编号查询订单明细表详细信息
    public List<OrdersItem> getOrdersItemByOrderId(String orderId);
    // 根据订单编号更新订单状态
```

```
public boolean updateOrdersState(Orders orders);
// 根据订单编号删除订单
public boolean deleteOrders(String orderId);}
```

2. 接口实现类设计

代码如下：

```
public class OrderDaoImpl implements IOrdersDao {
  /*
```

(1) 显示我的订单。

该方法用来查询表 orders 分页单位内商品记录功能，以登录用户名、请求页号和每页记录为参数，返回结果为商品集合 List<Orders>分页边界类型。代码如下：

```
*/
    @Override
 public List<Orders> getOrdersPageByUser(String username, int currentPage, int pageSize) {
     List<Orders> orders = new ArrayList<Orders>();
     String strSql = "select * from orders  where username=?  order by state asc limit ?,?";
     return (List<Orders>) DBUtil.executeQuery(strSql, new IResultSetUtil() {
         @Override
     public Object doHandler(ResultSet rs) throws SQLException {
        while (rs.next()) {
           Orders orders1 = new Orders();
           orders1.setUsername(rs.getString("username"));
           orders1.setTruename(rs.getString("truename"));
           orders1.setOrderId(rs.getString("orderId"));
           orders1.setPhone(rs.getString("phone"));
           orders1.setSum(rs.getFloat("sum"));
           orders1.setState(rs.getInt("state"));
           orders1.setAddtime(rs.getString("addtime"));
           orders.add(orders1);
           }
           return orders;
        }
     }, username, (currentPage - 1) * pageSize, pageSize);
}
/*
```

(2) 查询我的订单数目。

该方法用来查询表 orders 中属于登录用户的订单数量，返回结果为整数类型，表示记录数。代码如下：

```
*/
    @Override
  public int getOrdersTotalByUser(String username) {
        String strSql = "select count(*)  from orders  where username=?";
```

```
        Object obj = DBUtil.executeQuery(strSql, username);
        return Integer.parseInt(obj.toString());
    }
/*
```

(3) 根据订单编号查询订单主表详细信息。

该方法用来查询订单主表(orders)某一订单的详细信息，返回结果为 Orders 类型，表示某一订单实体对象。具体代码如下：

```
*/
@Override
    public Orders getOrdersByOrderId(String orderId) {
        String strSql = "select * from orders  where orderId=?";
        Orders orders = new Orders();
        return (Orders) DBUtil.executeQuery(strSql, new IResultSetUtil()
{       @Override
            public Object doHandler(ResultSet rs) throws SQLException {
                if (rs.next()) {
                    orders.setOrderId(rs.getString("orderId"));
                    orders.setUsername(rs.getString("username"));
                    orders.setTruename(rs.getString("truename"));
                    orders.setPhone(rs.getString("phone"));
                    orders.setSum(rs.getFloat("sum"));
                    orders.setState(rs.getInt("state"));
                    orders.setAddtime(rs.getString("addtime"));
                    orders.setPostcode(rs.getString("postcode"));
                    orders.setAddress(rs.getString("address"));
                    return orders;
                }
                return null;
            }
        }, orderId);
    }
/*
```

(4) 根据订单编号查询订单明细表详细信息。

该方法用来查询订单明细表(ordersitem) 某一订单编号的详细信息，返回结果为 List<OrdersItem>集合类型，表示某一订单实体对象。具体代码如下：

```
*/
    @Override
    public List<OrdersItem> getOrdersItemByOrderId(String orderId) {
        List<OrdersItem> ordersItems = new ArrayList<OrdersItem>();
        String strSql = "select * from ordersitem where orderId=?";
        return (List<OrdersItem>) DBUtil.executeQuery(strSql, new
IResultSetUtil() {
            @Override
            public Object doHandler(ResultSet rs) throws SQLException {
                while (rs.next()) {
                    OrdersItem orders1 = new OrdersItem();
```

```
                orders1.setOrderId(rs.getString("orderId"));
                orders1.setPid(rs.getInt("pid"));
                orders1.setShopname(rs.getString("shopname"));
                orders1.setShopnum(rs.getInt("shopnum"));
                orders1.setPrice(rs.getFloat("price"));
                ordersItems.add(orders1);
            }
            return ordersItems;
        }
    }, orderId);
}
/*
```

(5) 更新订单状态方法。

该方法用来实现订单主表(orders)订单状态的功能，以订单实体对象(orders)为参数，返回结果为布尔类型，表示更新商品信息是否成功。代码如下：

```
*/
    @Override
    public boolean updateOrdersState(Orders orders) {
        String strSql = "update orders set state=? where orderId=?";
        return DBUtil.executeUpdate(strSql, orders.getState(), orders.getOrderId()) > 0;
    }
/*
```

(6) 删除订单方法。

该方法用来实现删除订单表(orders)一条记录的功能，以订单编号 orderId 为参数，返回结果为布尔类型，表示删除商品信息是否成功。代码如下：

```
*/
    @Override
    public boolean deleteOrders(String orderId) {
        String strSql = "delete FROM orders where orderId=?";
        return DBUtil.executeUpdate(strSql, orderId) > 0;
    }
```

(二)我的订单子模块逻辑层设计

1. 接口类设计

接口类中的方法与持久层一样，这里不再赘述。

2. 接口实现类设计

代码如下：

```
public class OrdersServiceImpl implements IOrdersService {
    @Override
```

```java
    public List<Orders> getOrdersPageByUser(String username, int currentPage, int pageSize) {
        IOrdersDao ordersDao=new OrderDaoImpl();
        return ordersDao.getOrdersPageByUser(username,currentPage,pageSize);
    }
    @Override
    public int getOrdersTotalByUser(String username) {
        IOrdersDao ordersDao=new OrderDaoImpl();
        return ordersDao.getOrdersTotalByUser(username);
    }
    @Override
    public Orders getOrdersByOrderId(String orderId) {
        IOrdersDao ordersDao=new OrderDaoImpl();
        return ordersDao.getOrdersByOrderId(orderId);
    }
    @Override
    public List<OrdersItem> getOrdersItemByOrderId(String orderId) {
        IOrdersDao ordersDao=new OrderDaoImpl();
        return ordersDao.getOrdersItemByOrderId(orderId);
    }
    @Override
    public boolean updateOrdersState(Orders orders) {
        IOrdersDao ordersDao=new OrderDaoImpl();
        return ordersDao.updateOrdersState(orders);
    }
    @Override
    public boolean deleteOrders(String orderId) {
        IOrdersDao ordersDao=new OrderDaoImpl();
        return ordersDao.deleteOrders(orderId);
    }
}
```

(三)我的订单子模块控制层设计

1. 显示我的订单信息功能设计

当用户登录成功后，进入个人中心，在 left.jsp 页面单击"我的订单"的链接时，系统就会交给 OrdersInfoServlet 来处理，该类通过调用业务逻辑类的方法 getOrdersPageByUser 来实现全部商品分页展示的功能。具体代码如下：

```java
@WebServlet(name = "OrdersInfoServlet",urlPatterns = "/ordersInfo.do")
public class OrdersInfoServlet extends HttpServlet {
    protected void doPost(HttpServletRequest req, HttpServletResponse resp)
            throws ServletException, IOException {doGet(req,resp);}
    protected void doGet(HttpServletRequest req, HttpServletResponse resp)
            throws ServletException, IOException {
        HttpSession session=req.getSession();
        if (session.getAttribute("user")!=null){
            User user1= (User) session.getAttribute("user");
            try{
                int currentPage = Integer.parseInt(req.getParameter("currentPage"));
                IOrdersService ordersService=new OrdersServiceImpl();
```

```
            int totalSize = ordersService.getOrdersTotalByUser(user1.getUsername());
            int totalPage=totalSize/ Tally.SHOP_PAGE_SIZE;
                if(totalSize%Tally.SHOP_PAGE_SIZE!=0)
                    totalPage++;
                if (currentPage < 1) {
                    currentPage = 1;
                }
                if (currentPage > totalPage) {
                    currentPage = totalPage;
                }
                Pager page = new Pager(currentPage, totalSize);
                List<Orders> ordersInfo=ordersService.getOrdersPageByUser
(user1.getUsername(),currentPage,Tally.SHOP_PAGE_SIZE);
                String url = req.getRequestURI() + "?" + req.getQueryString();
                int index = url.lastIndexOf("currentPage=");
                if (index != -1) {
                    url = url.substring(0, index);
                }
        req.setAttribute("ordersInfo",ordersInfo);
        req.setAttribute("page", page);
        req.setAttribute("url", url);
                req.getRequestDispatcher("myOrdersInfo.jsp").forward(req,
resp);} catch (NumberFormatException ex) {
        String msg="页码传输异常";
        req.setAttribute("msg",msg);
                req.getRequestDispatcher("result.jsp").forward(req,resp);}
            }
        else{
        resp.sendRedirect("login.jsp");
            }
        }
}
```

2. 显示我的订单详细信息功能设计

当用户登录成功后，进入个人中心，在 orders.jsp 页面单击每一个订单编号的链接时，系统就会交给 OrdersListServlet 来处理，该类通过调用业务逻辑类的方法 getOrdersItemByOrderId 生成订单主表实体类对象 orders 和订单详细表实体类对象，并传回 myOrdersList.jsp 页面。代码如下：

```
@WebServlet(name = "OrdersListServlet",urlPatterns = "/ordersList.do")
public class OrdersListServlet extends HttpServlet {
protected void doPost(HttpServletRequest req, HttpServletResponse resp)
throws ServletException, IOException { doGet(req,resp); }
protected void doGet(HttpServletRequest req, HttpServletResponse resp)
throws ServletException, IOException {
    String orderId=req.getParameter("orderId");
    IOrdersService ordersService=new OrdersServiceImpl();
    Orders  rders=ordersService.getOrdersByOrderId(orderId);
```

```
    List<OrdersItem>
ordersItemList=ordersService.getOrdersItemByOrderId(orderId);
    req.setAttribute("orders",orders);
   req.setAttribute("ordersItemList",ordersItemList);
req.getRequestDispatcher("myOrdersList.jsp").forward(req,resp);
      }
}
```

3. 修改我的订单状态功能设计

当用户登录成功后，进入个人中心，在 myOrdersList.jsp 页面当用户发出修改订单状态的请求后，系统就会交给 UpdateOrderStateServlet 来处理。具体代码如下：

```
@WebServlet(name = "UpdateOrderStateServlet",urlPatterns = "/updateOrderState.do")
public class UpdateOrderStateServlet extends HttpServlet {
protected void doPost(HttpServletRequest req, HttpServletResponse resp)
throws ServletException, IOException {
HttpSession session = req.getSession();
User user1= (User) session.getAttribute("user");
String orderId = req.getParameter("orderId");
String states = req.getParameter("state");
if (states==null ||"".equals(states)){
 String msg = "订单状态未做任何改变";
   req.setAttribute("msg", msg);
   req.getRequestDispatcher("myresult.jsp").forward(req, resp);
        }
    int state=Integer.parseInt(states);
    Orders orders = new Orders();
    orders.setOrderId(orderId);
    orders.setState(state);
    float sum = Float.parseFloat(req.getParameter("sum"));
    int mypoint=(int)sum;
    IOrdersService ordersService = new OrdersServiceImpl();
    if (ordersService.updateOrdersState(orders)) {
      if (state == 2) {
      IUserService userService = new UserServiceImpl();
      Point point = new Point();
      point.setUsername(user1.getUsername());
      point.setOrderid(orderId);
               point.setPointtime(StringHelper.getCurrentFormatDate());
      point.setMypoint(mypoint);
      point.setDescription("购物赠送积分");
      userService.addPoint(point);
      String msg = "修改订单状态成功,并赠送积分" + mypoint;
       req.setAttribute("msg", msg);
               req.getRequestDispatcher("myresult.jsp").forward(req, resp);}
        else{ String msg = "修改订单状态成功!";
      req.setAttribute("msg", msg);
               req.getRequestDispatcher("myresult.jsp").forward(req, resp);
```

```
            }       }
        else{ String msg="修改订单状态未成功";
        req.setAttribute("msg",msg);
req.getRequestDispatcher("myresult.jsp").forward(req,resp);
            }
        }
    protected void doGet(HttpServletRequest req, HttpServletResponse resp)
throws ServletException, IOException {
        doPost(req,resp);    }
}
```

【代码说明】

(1) 运用 req 对象接收订单状态和订单编号,然后封装到订单实体类对象(orders)中。

(2) 用户修改状态的请求有两种,由订单状态 1(未付款订单)改为 2(已付款订单),由订单状态 3(已发货)改为 4(已收货)。

当用户进行第一种请求时,调用业务逻辑类 updateOrdersState(orders)方法,进行订单状态更新,如果返回值为真,还要赠送用户相应的积分,积分总额等于购物总额的整数。还要调用积分业务逻辑类方法 addPoint(point)并把赠送积分记录插入到 point(积分)表里。

4. 删除我的订单功能设计

当用户登录成功后,进入个人中心,在 orders.jsp 页面发出删除订单的请求后,系统就会交给 OrdersDeleServlet 来处理。具体代码如下:

```
@WebServlet(name = "OrdersDeleServlet",urlPatterns = "/ordersDele.do")
public class OrdersDeleServlet extends HttpServlet {
protected void doPost(HttpServletRequest req, HttpServletResponse resp)
throws ServletException, IOException { doGet(req, resp); }
    protected void doGet(HttpServletRequest req, HttpServletResponse resp)
throws ServletException, IOException {
        String orderId =req.getParameter("orderId");
        IOrdersService ordersService = new OrdersServiceImpl();
        Orders orders=new Orders();
        orders=ordersService.getOrdersByOrderId(orderId );
        if (orders.getState()!=1){
        String msg = "该订单状态已付款,不能撤销";
        req.setAttribute("msg", msg);
           req.getRequestDispatcher("result.jsp").forward(req, resp);
        } else {
        if (ordersService.deleteOrders(orderId)) {
        String msg = "成功删除!";
        req.setAttribute("msg", msg);
            req.getRequestDispatcher("result.jsp").forward(req, resp);
          } else {
         String msg = "删除失败!";
         req.setAttribute("msg", msg);
            req.getRequestDispatcher("result.jsp").forward(req, resp);
           }
        }
```

            }
    }

【代码说明】

删除订单时，首先看一下订单状态，如果订单状态不是 1，说明该订单已经处理，则不能删除，否则订单业务处理类方法由 ordersService.deleteOrders(orderId)来删除订单。

> **小提示**：Servlet 与 Java 类的区别
> Servlet 是一个供其他 Java 程序(Servlet 引擎)调用的 Java 类，它不能独立运行，它的运行完全由 Servlet 引擎来控制和调度。
> 针对客户端的多次 Servlet 请求，通常情况下，服务器只会创建一个 Servlet 实例对象。也就是说，Servlet 实例对象一旦创建，它就会驻留在内存中，为后续的其他请求服务，直至 Web 容器退出，Servlet 实例对象才会销毁。
> 在 Servlet 的整个生命周期内，Servlet 的 init()方法只被调用一次。而对一个 Servlet 的每次访问请求都导致 Servlet 引擎调用一次 Servlet 的 service()方法。对于每次访问请求，Servlet 引擎都会创建一个新的 HttpServletRequest 请求对象和一个新的 HttpServletResponse 响应对象，然后将这两个对象作为参数传递给它调用的 Servlet 的 service()方法，service()方法再根据请求方式分别调用 doXXX()方法。
> 如果在<servlet>元素中配置了一个<load-on-startup>元素，那么 Web 应用程序在启动时，就会装载并创建 Servlet 的实例对象，以及调用 Servlet 实例对象的 init()方法。

(四)我的订单子模块视图层设计

1. 显示我的订单(myOrdersInfo.jsp)页面设计

代码如下：

```jsp
<%@ taglib prefix="c" uri="http://java.sun.com/jsp/jstl/core" %>
<%@ page contentType="text/html;charset=UTF-8" language="java" %>
<!DOCTYPE html>
<html>
<head>
    <meta charset="utf-8" />
    <meta http-equiv="X-UA-Compatible" content="ie=edge">
    <title>个人中心</title>
 <link type="text/css" rel="stylesheet" href="css/font-awesome.css">
    <link type="text/css" rel="stylesheet" href="css/font-awesome.min.css">
    <link type="text/css" rel="stylesheet" href="css/reset.css">
    <link type="text/css" rel="stylesheet" href="css/ordersInfo.css">
   </head>
<body>
<jsp:include page="head.jsp"/>
<div class="crumb">
    <div class="w">
        <div class="crumb-con">
```

```
                <a href="index.do" class="link">MMAIL</a>
                <span>></span>
                <span class="link-text">我的订单</span>
            </div>
        </div>
</div>
<div class="page-wrap w">
    <jsp:include page="left.jsp"/>
    <div class="content with-nav">
        <div class="panel">
            <div class="panel-title">我的订单</div> <span id="span12">${resu}</span>
            <div class="panel-body">
                <table class="table" cellspacing="0" cellpadding="0">
                    <thead>
                    <tr>
                        <th width="15%">订单编号</th>
                        <th width="15%">真实姓名</th>
                        <th width="10%">手机</th>
                        <th width="10%">金额</th>
                        <th width="15%">时间</th>
                        <th width="10%">状态</th>
                        <th width="25%">操作</th>
                    </tr>
                    </thead>
                    <tbody>
                    <c:forEach items="${ordersInfo}" var="orders">
                        <tr>
                            <td width="15%">${orders.orderId}</td>
                            <td width="15%">${orders.truename}</td>
                            <td width="15%">${orders.phone}</td>
                            <td width="10%">${orders.sum}</td>
                            <td width="15%">${orders.addtime}</td>
                            <td width="10%">
                        <c:if test="${orders.state==1}">未付款</c:if>
                        <c:if test="${orders.state==2}">已付款</c:if>
                        <c:if test="${orders.state==3}">已发货</c:if>
                        <c:if test="${orders.state==4}">订单完成</c:if>
                            </td>
                            <td width="20%"><a class="link" href="ordersDele.do?orderId=${orders.orderId}" onClick="return confirm('您确定进行撤销操作吗？')">撤销</a>
                            <a class="link" href="ordersList.do?orderId=${orders.orderId}">订单明细</a></td>
                        </tr>
                    </c:forEach>
                    </tbody>
                </table>
            </div>
        </div>
```

```html
        <div class="pg-content">
            <form class="page-form" method="post" onsubmit="return checkForm();" action="${url}">
            <input type="text" name="currentPage" id="currengPage" class="page" >
                <input type="submit" value="go" >
            </form>
            <c:if test="${page.hasFirst}">
                <li><a href="${url}currentPage=1">首页</a></li>
            </c:if>
            <c:if test="${page.hasPrevious}">
                <li><a href="${url}currentPage=
                    ${page.currentPage-1}">上一页</a></li>
            </c:if>
            <c:if test="${page.hasNext}">
            <li><a href="${url}currentPage=
                    ${page.currentPage+1}">下一页</a></li>
            </c:if>
            <c:if test="${page.hasLast}">
            <li><a href="${url}currentPage=
                ${page.totalPage}">尾页</a></li>
            </c:if>
        <span class="pg-total">
            当前第${page.currentPage}页，总共${page.totalPage}页 </span>
        </div>
    </div>
    <jsp:include page="foot.jsp"/>
    </body>
</html>
```

【代码说明】

该页面显示登录用户的所有订单，并进行分页显示，该页面接收 AdminAllOrdersServlet 传回的分页集合对象 ordersInfo、分页容器对象 page 及地址栏变量 url，通过 JSTL 标签进行遍历解析，把内容显示到当前页面上。运行效果如图 7-4 所示。

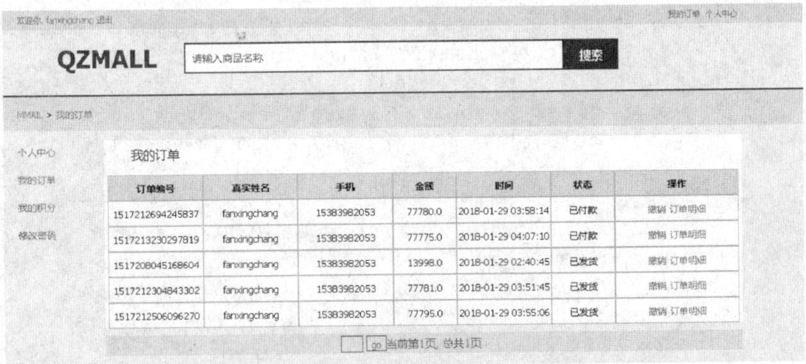

图 7-4 我的订单(myOrdersInfo.jsp)页面

2. 显示我的订单明细(myOrdersList.jsp)页面设计

代码如下：

```
<%@ taglib prefix="c" uri="http://java.sun.com/jsp/jstl/core" %>
<%@ page contentType="text/html;charset=UTF-8" language="java" %>
<!DOCTYPE html>
<html>
<head>
    <meta charset="utf-8" />
    <meta http-equiv="X-UA-Compatible" content="ie=edge">
    <title>个人中心</title>
    <link type="text/css" rel="stylesheet" href="css/font-awesome.css">
    <link type="text/css" rel="stylesheet" href="css/font-awesome.min.css">
    <link type="text/css" rel="stylesheet" href="css/reset.css">
    <link type="text/css" rel="stylesheet" href="css/ordersList.css">
</head>
<body>
<jsp:include page="head.jsp"/>
<div class="crumb">
    <div class="w">
        <div class="crumb-con">
            <a href="index.do" class="link">MMAIL</a>
            <span></span>
            <span class="link-text">我的订单</span>
        </div>
    </div>
</div>
<div class="page-wrap w">
    <jsp:include page="left.jsp"/>
    <div class="content with-nav">
        <div class="panel">
            <div class="panel-title">修改订单状态</div> <span id="span12">${resu}</span>
            <form name="form1" method="post" action="updateOrderState.do">
            <div class="panel-body">
                <div class="form-line">
                    <span class="label">收货人：</span>
                    <span class="text">${orders.truename}</span>
                    <input id="orderId" name="orderId" value="${orders.orderId}" type="hidden" readOnly="true">
                    <span class="label">手  机：</span>
                    <span class="text">${orders.phone}</span>
                </div>
                <div class="form-line">
                    <span class="label">地    址：</span>
                    <span class="text">${orders.address}</span>
                    <span class="label">邮  编：</span>
                    <span class="text">${orders.postcode}</span>
                </div>
                <div class="form-line">
                    <span class="label">金    额：</span>
                    <span class="text"><input type="text" id="sum" name="sum" value="${orders.sum}" readOnly="true"></span>
```

```html
            <span class="label">状  态：</span>
              <span class="text"><select name="state">
                <c:if test="${orders.state==1}">
                 <option value="1" selected="selected" disabled="disabled">未付款</option>
                 <option value="2" >已付款</option>
                </c:if>
              <c:if test="${orders.state==2}">
                <option value="2" selected="selected" disabled="disabled">客户已付款</option>
                </c:if>
                <c:if test="${orders.state==3}">
                <option value="3" selected="selected" disabled="disabled">商家已发货</option>
                <option value="4" >用户已收货</option>
                </c:if>
                <c:if test="${orders.state==4}">
                <option value="4" selected="selected" disabled="disabled">用户已收货</option>
                </c:if>
                </select></span>
            </div>
            <div class="form-line">
              <span class="text"><input class="btn" type="submit" value="提交"></span>
            </div>
          </div>
        </form>
        <div class="panel-body">
          <table class="table" cellspacing="0" cellpadding="0">
            <thead>
            <tr>
                <th width="20%">订单编号</th>
                <th width="20%">商品编号</th>
                <th width="20%">商品名称</th>
                <th width="20%">数量</th>
                <th width="20%">价格</th>
            </tr>
            </thead>
            <tbody>
            <c:forEach items="${ordersItemList}" var="item">
                <tr>
                    <td width="20%">${item.orderId}</td>
                    <td width="20%">${item.pid}</td>
                    <td width="20%">${item.shopname}</td>
                    <td width="20%">${item.shopnum}</td>
                    <td width="20%">${item.price}</td>
                </tr>
            </c:forEach>
            </tbody>
          </table>
        </div>
      </div>
    </div>
    <jsp:include page="foot.jsp"/>
```

```
</body>
</html>
```

【代码说明】

该页面接收 OrdersListServlet 传回的订单主表实体类对象 order 和订单明细集合类 ordersItemList，通过 EL 和 JSTL 标签进行遍历解析，把某一订单编号的内容显示到当前页面上，运行效果如图 7-5 所示。

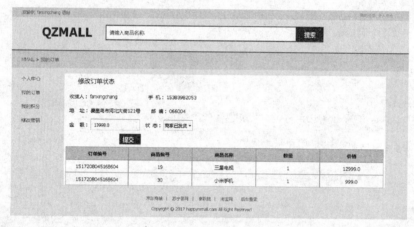

图 7-5　我的订单明细(myOrdersList.jsp)页面

3. 个人中心操作通用提示页(myresult.jsp)设计

代码如下：

```jsp
<%@ taglib prefix="c" uri="http://java.sun.com/jsp/jstl/core" %>
<%@ page contentType="text/html;charset=UTF-8" language="java" %>
<!DOCTYPE html>
<html>
<head>
    <meta charset="utf-8" />
    <meta http-equiv="X-UA-Compatible" content="ie=edge">
    <title>个人中心</title>
    <link type="text/css" rel="stylesheet" href="css/font-awesome.css">
    <link type="text/css" rel="stylesheet" href="css/font-awesome.min.css">
    <link type="text/css" rel="stylesheet" href="css/reset.css">
    <link type="text/css" rel="stylesheet" href="css/myresult.css">
</head>
<body>
<jsp:include page="head.jsp"/>
<div class="crumb">
    <div class="w">
        <div class="crumb-con">
            <a href="index.do" class="link">MMAIL</a>
            <span></span>
            <span class="link-text">个人中心</span>
        </div>
    </div>
```

```
</div>
<div class="page-wrap w">
   <jsp:include page="left.jsp"/>
   <div class="content with-nav">
      <div class="result-con">
      <h1 class="result-title">${msg}</h1>
      <div class="result-content">
      <a href="ordersInfo.do?currentPage=1" class="link">我的订单</a>
         </div>
         </div>
   </div>
   <jsp:include page="foot.jsp"/>
</body>
</html>
```

接着上一个页面，当修改订单状态并提交时，显示的结果如图 7-6 所示，表示修改订单状态成功。

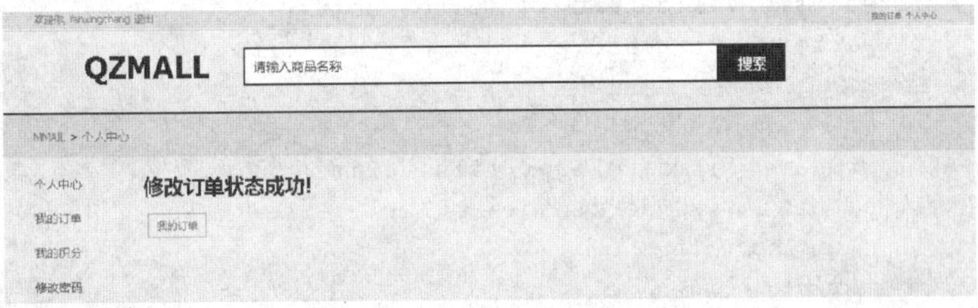

图 7-6　显示订单状态修改成功

(五)配置 web.xml

代码如下：

```xml
<?xml version="1.0" encoding="UTF-8"?>
<web-app xmlns="http://xmlns.jcp.org/xml/ns/javaee"
 xmlns:xsi="http://www.w3.org/2001/XMLSchema-instance"
 xsi:schemaLocation="http://xmlns.jcp.org/xml/ns/javaee
http://xmlns.jcp.org/xml/ns/javaee/web-app_3_1.xsd"
         version="3.1">
   <display-name>ecshop</display-name>
   <welcome-file-list>
      <welcome-file>index.jsp</welcome-file>
   </welcome-file-list>
   <context-param>
      <param-name>login_page</param-name>
      <param-value>/login.jsp</param-value>
   </context-param>
   <context-param>
      <param-name>login_admin</param-name>
      <param-value>/adminlogin.jsp</param-value>
   </context-param>
   <filter>
```

```xml
        <filter-name>validateLogn</filter-name>
        <filter-class>com.qzmall.filter.ValidateFilter</filter-class>
    </filter>
    <filter-mapping>
        <filter-name>validateLogn</filter-name>
        <url-pattern>/cart.jsp</url-pattern>
    </filter-mapping>
    <filter-mapping>
        <filter-name>validateLogn</filter-name>
        <url-pattern>/cartServlet.jsp</url-pattern>
    </filter-mapping>
    <filter-mapping>
        <filter-name>validateLogn</filter-name>
        <url-pattern>/userInfoSave.do</url-pattern>
    </filter-mapping>
    <filter-mapping>
        <filter-name>validateLogn</filter-name>
        <url-pattern>/userInfoSave.do</url-pattern>
    </filter-mapping>
    <filter-mapping>
        <filter-name>validateLogn</filter-name>
        <url-pattern>/myOrdersInfo.jsp</url-pattern>
    </filter-mapping>
    <filter-mapping>
        <filter-name>validateLogn</filter-name>
        <url-pattern>/myOrdersInfo.jsp</url-pattern>
    </filter-mapping>
    <filter-mapping>
        <filter-name>validateLogn</filter-name>
        <url-pattern>/ordersInfo.do</url-pattern>
    </filter-mapping>
    <filter-mapping>
        <filter-name>validateLogn</filter-name>
        <url-pattern>/ordersList.do</url-pattern>
    </filter-mapping>
    <filter-mapping>
        <filter-name>validateLogn</filter-name>
        <url-pattern>/ordersDele.do</url-pattern>
    </filter-mapping>
    <filter-mapping>
    <filter-name>validateLogn</filter-name>
    <url-pattern>/modifypass.jsp</url-pattern>
    </filter-mapping>
    <filter-mapping>
        <filter-name>validateLogn</filter-name>
        <url-pattern>/modifypass.jsp</url-pattern>
    </filter-mapping>
    <filter-mapping>
```

```xml
            <filter-name>validateLogn</filter-name>
            <url-pattern>/modifyPass.do</url-pattern>
    <filter-mapping>
            <filter-name>validateLogn</filter-name>
            <url-pattern>/modifyPass.do</url-pattern>
    </filter-mapping>
    <filter-mapping>
            <filter-name>validateLogn</filter-name>
            <url-pattern>/myPoint.do</url-pattern>
    </filter-mapping>
    <filter-mapping>
            <filter-name>validateLogn</filter-name>
            <url-pattern>/myPoint.jsp</url-pattern>
    </filter-mapping>
    <filter-mapping>
    <filter-name>validateLogn</filter-name>
    <url-pattern>/order.jsp</url-pattern>
     </filter-mapping>
    <filter-mapping>
            <filter-name>validateLogn</filter-name>
            <url-pattern>/orderServlet.do</url-pattern>
    </filter-mapping>
    <filter-mapping>
            <filter-name>validateLogn</filter-name>
            <url-pattern>/checkout.jsp</url-pattern>
    </filter-mapping>
    <filter-mapping>
            <filter-name>validateLogn</filter-name>
            <url-pattern>/checkOut.do</url-pattern>
    </filter-mapping>
    <filter-mapping>
            <filter-name>validateLogn</filter-name>
            <url-pattern>/updateOrderState.do</url-pattern>
    </filter-mapping>
</web-app>
```

【代码说明】

上述页面及控制层的 servlet(包括第六章)讲的内容都必须是登录以后才能进行操作，所以 web.xml 配置了一个登录过滤器，从而限制未登录用户的非法操作。

课堂技能训练：

【实训操作内容】完成我的订单状态修改功能设计。

【实训操作步骤】

(1) 持久层设计。

(2) 业务逻辑层设计。

(3) 控制层设计(重点，订单状态修改了，应奖励用户积分)。

(4) 视图层设计。

【知识拓展】

订单是在用户提交购物车里购买的商品时保存的订单信息，为初始信息，此时状态为未支付。但本网站并未实现网站的在线支付功能，因为要实现在线支付功能，必须通过中间公司间接与银行对接，所以这部分功能还不能实现。要实现在线支付的功能一般是需要用户选择支付方式，进行相应的付款操作。此时订单状态需要根据支付结果来修改，如果支付成功，银联或者支付宝、微信会从后台通知到你的应用，订单状态修改为已支付；如果没有支付成功，订单状态还是未支付，网站需要和银联或者支付宝、微信等第三方支付机构建立对账机制，保证程序内出现订单状态不一致时能够及时解决。

课后训练

一、选择题

1. 在 JavaEE 中，Servlet 是在服务器端运行，以处理客户端请求而做出的响应的程序，下列选项中属于 Servlet 生命周期阶段的是（　　）。

　　A. 加载和实例化　　　　B. 初始化　　　　　　C. 服务
　　D. 销毁　　　　　　　　E. 以上全部

2. 在 JavaEE 中的 MVC 设计模式中，（　　）负责接收客户端的请求数据。

　　A. JavaBean　　　　B. JSP　　　　C. Servlet　　　　D. HTML

3. 过滤器应实现的接口是（　　）。

　　A. HttpServlet　　　B. HttpFilter　　　C. ServletFilter　　　D. Filter

4. 开发 Java Web 应用程序的时候，创建一个 Servlet，该 Servlet 重写了父类的 doGet() 和 doPost() 方法，那么其父类可能是（　　）。

　　A. RequestDispatcher　　　　　　B. HttpServletResponse
　　C. HttpServletRequest　　　　　　D. HttpServlet

5. 在 Java Web 开发中，如果某个数据需要跨多个请求存在，则数据应该存储在（　　）中。

　　A. session　　　　B. page　　　　C. request　　　　D. response

6. 在开发 Java Web 应用程序的时候，HTTP 请求消息使用 Get 或 POST 方法以便在 Web 上传输数据，下列关于 GET 和 POST 方法描述正确的是（　　）。

　　A. POST 请求的数据在地址栏不可见
　　B. GET 请求提交的数据在理论上没有长度限制
　　C. POST 请求对发送的数据的长度限制在 240~255 个字符
　　D. GET 请求提交数据更加安全

7. 在 JSP 中有 EL 表达式 ${10*10 ne 10}，结果是（　　）。

　　A. 100　　　　B. true　　　　C. false　　　　D. 以上都不对

8. 关于JSTL标签的分类以下说法正确的是(　　)。
   A. 通用标签与迭代标签　　　　B. 核心标签与迭代标签
   C. 核心标签与SQL标签　　　　D. 以上都不是

9. 在Java Servlet API中，HttpServletRequest接口的(　　)方法用于返回当前请求相关联的会话，如果没有，返回null。
   A. getSession() //=getSession(true)默认为：getSession(true)
   B. getSession(true)
   C. getSession(false)
   D. getSession(null)

10. 在Java Web开发中，不同的客户端需要共享数据，则数据应该存储在(　　)中。
    A. session　　　B. application　　　C. request　　　D. response

11. 利用三层结构搭建某网上书店系统，设计用户登录界面，如果你是设计人员，你将在三层结构的(　　)中实现。
    A. 模型层　　　B. 表示层　　　C. 数据访问层　　　D. 业务逻辑层

12. 在JavaEE中，(　　)接口定义了getSession()方法。
    A. httpServlet　　　　　　　　B. HttpSession
    C. HttpServletRequest　　　　D. HttpServletResponse

13. JSP标准标签库(JSTL)提供的主要标签库中，其中(　　)可用于操作数据库。
    A. 核心标签库　　　　　　　　B. I18N与格式化标签库
    C. XML标签库　　　　　　　　D. SQL标签库

14. JSP表达式可用于在网页上生成动态的内容并代替JSP元素，JSP表达式的语法是(　　)。
    A. {EL expression}　　　　　B. ${EL expression}
    C. @{EL expression}　　　　D. #{EL expression}

15. 在J2EE的体系结构中，系统的业务和功能代码组成了三层结构中的(　　)。
    A. 表示层　　　B. 中间层　　　C. 数据层　　　D. 客户端层

16. 在JSTL的迭代标签<forEach>的属性中，用于指定要遍历的对象集合(　　)。
    A. var　　　B. items　　　C. value　　　D. varStatus

17. 如果不希望JSP网页支持session，应该(　　)。
    A. 调用HttpSession的invalidate()方法
    B. 设置<%@ page session="false">
    C. 在JSP页面中写入代码HttpSession session=request.getSession(false);
    D. 调用HttpSession的setMaxInactiveInterval()，将时间设置为0

18. Servlet是一个在服务器上运行、处理请求信息并将其发送到客户端的Java程序，(　　)接受服务请求后，找到Servlet类、加载Servlet类并创建Servlet对象。
    A. 客户端　　　B. IE浏览器　　　C. Servlet容器　　　D. Servlet本身

19. MVC 设计模式包含模型层、视图层和控制层，在下列组件中扮演控制器角色的是( )。

　　A. Servlet　　　　B. JSP　　　　C. JavaBean　　　　D. HTML

二、实际操作题

利用三层架构，写出用户模块中的个人积分浏览和修改密码的功能。

第七章 购物车及定单模块设计.pptx

第七章 习题答案.docx

第八章

后台维护模块设计

知识能力目标

1. 掌握组件常用的类与方法。
2. 掌握在 Servlet 中应用组件进行文件上传的方法。
3. 掌握商品添加和维护的基本流程。
4. 掌握后台订单维护的基本流程。
5. 能够运用上传组件进行文件的上传。
6. 能够完成商品添加和维护的编码设计。
7. 能够完成后台订单维护的编码设计。

问题提示

一个完整的电商平台由前台(客户端)和后台(服务端)两部分组成。

前台(客户端)按目前主流平台分为 Web、WAP、iOS 和 Android；前台主要面向的对象是广大互联网用户，作用是提供一个让用户从浏览、挑选商品到下单付款全过程的平台，所以搭建前台的重点是建立和打通整个购物流程，在这之后再提供因为业务需要不断添加的功能点和提升用户体验的交互。

后台(服务端)的面向对象是公司内部人员，主要作用是为前台业务提供支持并对数据结构、储存和流向进行控制的一套体系。

如果将电商的整套体系比作一座大 house 的话，那么后台(服务端)就是埋在地下的地基，而前台(客户端)则是建在地面上的 house。用户来到 house 后不会关心地基是怎么搭建的，他只会注意到 house 的样式和 house 中的房间是什么样的；前后台也像这座 house 的地下和地上部分一样有着相辅相成的关系。

问题：电子商务网站的后台主要提供哪些服务？

第一节　添加商品子模块设计

一、文件上传组件 commons-fileUpload 简介

commons-fileUpload 上传组件是 Apache 的一个开源项目，该组件需要 commons-io 包的支持。它使用方便，同样可以实现一个或多个文件的上传，也可实现限制上传文件大小等功能。

在文件上传中，文件上传请求由有序的表单项列表组成，fileUpload 能够解析上传请求，然后响应提供一个单独的文件表单项的列表。每个这样的表单项都实现了 FileItem 接口，并包含文件名等属性。

(一)使用要求

(1) 使用 commons-fileUpload 组件上传文件时，需要将 form 表单的 enctype 属性设置

为 multipart/form-data；同时还需要设置上传文件在内存中的大小，多余的部分存储在磁盘中。

(2) 将相应的 commons-fileUpload 包和 commons-io 包复制放入到用户的 Web 工程 WEN-INF/lib 目录下即可。

(二)使用步骤

(1) 创建磁盘工厂 DiskFileItemFactory 对象，用来配置上传组件 ServletFileUpload。代码如下：

```
DiskFileItemFactory factory = new DiskFileItemFactory();
```

DiskFileItemFactory 类的常用方法，如表 8-1 所示。

表 8-1 DiskFileItemFactory 类的常用方法

方 法	返回值	描 述
setSizeThreshold()	void	该方法需要传入一个 int 型参数，用来设置最大内存大小(Byte 为单位)
setRepositoryPath()	void	该方法需要传入一个 string 型参数，用来设置临时文件目录
getRepository()	File	获取保存临时文件地址

(2) 创建 ServletFileUpload 实例，即创建上传文件的句柄。

可通过 DiskFileItemFactory 实例构造 ServletFileUpload 对象。代码如下：

```
ServletFileUpload upload = newServletFileUpload(factory);
```

ServletFileUpload 类的常用方法，如表 8-2 所示。

表 8-2 ServletFileUpload 类的常用方法

方 法	返回值	描 述
isMultipartContent()	boolean	检查是否是一个文件上传请求
parseRequest()	List	从 request 对象中得到所有上传域表单

二、添加商品功能实现

(一)添加商品持久层设计

1. 接口类设计

在 com.qzmall.dao.IShopDao 类下添加以下方法：

```java
public interface IShopDao {
    //添加商品
    public boolean AddShop(Shop shop);
}
```

2. 接口实现类设计

在 com.qzmall.dao.impl.ShopDaoImpl 类下实现商品添加方法：

```
public class ShopDaoImpl implements IShopDao {
 /*
```

该方法用来向表 shop 添加一条记录,以商品对象为参数,返回结果为布尔类型,表示商品记录是否插入成功。代码如下:

```
*/
@Override
public boolean AddShop(Shop shop) {
 String strSql = "insert into shop
(cid,sid,shopname,shopinfo,price,stock,
shopdate,image1,image2,image3,description) values (?,?,?,?,?,?,?,?,?,?,?)";
    return DBUtil.executeUpdate(strSql, shop.getCategory().getCid(),
shop.getSubCate().getSid(), shop.getShopname(), shop.getShopinfo(),
shop.getPrice(), shop.getStock(), shop.getShopdate(), shop.getImage1(),
shop.getImage2(), shop.getImage3(), shop.getDescription()) > 0;
   }
 }
```

(二)商品添加子模块逻辑层设计

1. 接口类设计

接口类放在 com.qzmall.service.IShopService 类下,接口类设计的方法与持久层相同,这里不再赘述。

2. 接口实现类设计

接口类方法放在 com.qzmall.service.impl.ShopServiceImpl 类下。代码如下:

```
public class ShopServiceImpl implements IShopService {
    @Override
public boolean AddShop(Shop shop) {
    IShopDao shopDao=new ShopDaoImpl();
    return shopDao.AddShop(shop);
   }
}
```

(三)上传商品工具类(UploadUtils)——生成唯一文件方法设计

代码如下:

```
public class UploadUtils {
    /**
     * 生成唯一的文件名:
     */
    public static String getUUIDFileName(String fileName){
        //将文件名的前面部分进行截取:xx.jpg   --->   .jpg
        int idx = fileName.lastIndexOf(".");
        String extention = fileName.substring(idx);
        String uuidFileName = UUID.randomUUID().toString().replace("-",
"")+extention;
```

```
        return uuidFileName;
    }
    public static void main(String[] args) {
        System.out.println(getUUIDFileName("1.jpg"));
    }
}
```

(四)商品添加子模块控制层类设计

商品管理模块控制层类在 com.qzmall.servlet.shop 包下。

1. 跳转到增加商品页面(addshop.jsp)

当管理员从 menu.jsp 页面发出添加商品的请求时,系统就会交给 AddShopServlet 来处理,该类通过调用业务逻辑类的方法 getShopPage 来实现全部商品分页展示的功能。具体代码如下:

```
@WebServlet(name = "AddShopServlet",urlPatterns = "/admin/addshop.do")
public class AddShopServlet extends HttpServlet {
protected void doGet(HttpServletRequest req, HttpServletResponse resp)
throws ServletException, IOException {
    String con=req.getParameter("con");
    req.setAttribute("con",con);
    IShopService categoryService=new ShopServiceImpl();
    ArrayList<Category> categoryArrayList=categoryService.getCategoryAll();
    req.setAttribute("categoryArrayList",categoryArrayList);
    req.getRequestDispatcher("addshop.jsp").forward(req,resp);
    }
}
```

【代码说明】

在上述代码中,首先调用 categoryService.getCategoryAll() 得到所有分类对象 categoryArrayList,然后把它传回页面主页面 addshop.jsp,主要用于管理者在添加商品时,方便选择商品类别。

2. 保存添加商品信息功能设计

当管理员通过 addshop.jsp 页面发出商品提交信息时,就会交给 SaveShopServlet 类进行处理,通过调用商品业务逻辑类 AddShop 来实现商品添加的功能。具体代码如下:

```
@WebServlet(name = "SaveShopServlet",urlPatterns = "/admin/saveShop.do")
public class SaveShopServlet extends HttpServlet {
public void doPost(HttpServletRequest req, HttpServletResponse resp)
throws IOException, ServletException {
    Shop shop = new Shop();
    ICategoryService categoryService=new CategoryServiceImpl();
    int i=0;
    String []pic=new String[3];
    // 定义一个 Map 集合用于保存接收到的数据
    Map<String, String> map = new HashMap<String, String>();
```

```java
        // 1.创建一个磁盘文件项工厂对象
            DiskFileItemFactory diskFileItemFactory = new 
DiskFileItemFactory();
        // 2.创建一个核心解析类
            ServletFileUpload servletFileUpload = new 
            ServletFileUpload(diskFileItemFactory);
    // 3.解析 request 请求，返回的是 List 集合，List 集合中存放的是 FileItem 对象
        servletFileUpload.setSizeMax(1000*1024);
        List<FileItem> list=null;
        try {
            list = servletFileUpload.parseRequest(req);
        } catch (FileUploadException e) {
            // 如果出现这个异步，说明单个文件超出了 80KB
            error("上传的文件超出了1000KB", req, resp);
            return;
        }
        String url = null;
        for (FileItem fileItem : list) {
            // 判断是表单项还是文件上传项
            if (fileItem.isFormField()) {
            // 获得表单项的 name 属性的值
            String name = fileItem.getFieldName();
            // 获得表单项的值
             String value = fileItem.getString("UTF-8");
                         map.put(name, value);
            } else {
            // 如果是文件上传项
            // 获得文件上传的名称
        String fileName = fileItem.getName();
        if(fileName !=null && !"".equals(fileName)){
        if((!fileName.toLowerCase().endsWith(".jpg")) && 
(!fileName.toLowerCase().endsWith(".png")) ) {
            error("上传的图片扩展名必须是 jpg 或者 png", req, resp);
         return; }
                // 通过工具类获得唯一文件名
            String uuidFileName = 
UploadUtils.getUUIDFileName(fileName);
                // 获得文件上传的数据
            InputStream is = fileItem.getInputStream();
                // 获得文件上传的路径
        String path = 
this.getServletContext().getRealPath("/img//pic");
                // 将输入流对接到输出流就可以了
        url = path+"\\"+uuidFileName;
        pic[i]=uuidFileName;
        i++;
        System.out.println("url="+url);
        OutputStream os = new  FileOutputStream(url);
        int len = 0;
        byte[] b = new byte[1024];
```

```java
            while((len = is.read(b))!=-1){
                os.write(b, 0, len);
                }
            is.close();
            os.close();
             }
             }
            }
            // 获得ServletContext对象
            System.out.println(pic[1]);
            String ss = map.get("sid");
            int sid = Integer.parseInt(ss);
            int cid = categoryService.getCid(sid);
            Category category = new Category();
            category.setCid(cid);
            SubCate subCate = new SubCate();
            subCate.setSid(sid);
            shop.setShopname(map.get("shopname"));
            shop.setShopinfo(map.get("shopinfo"));
            shop.setCategory(category);
            shop.setSubCate(subCate);
                shop.setPrice(Float.parseFloat(map.get("price")));
shop.setShopdate(StringHelper.getCurrentFormatDate());
            shop.setStock(0);
            shop.setImage1(pic[0]);
            shop.setImage2(pic[1]);
            shop.setImage3(pic[2]);
        shop.setDescription(map.get("description"));
        // 将注册用户的信息存入到List集合中
        IShopService shopService = new ShopServiceImpl();
        if (shopService.AddShop(shop)) {
        String success = "商品添加成功！";
        this.getServletContext().setAttribute("msg", success);
         // 跳转到result.jsp，提示添加商品成功
                req.getRequestDispatcher("result.jsp").forward(req, resp);
            } else {
        String success = "商品添加失败！";
        this.getServletContext().setAttribute("msg", success);
            // 跳转到result.jsp，提示添加商品失败
        req.getRequestDispatcher("result.jsp").forward(req, resp);
            }
        }
    private void error(String msg, HttpServletRequest request,
HttpServletResponse response)
        throws ServletException, IOException {
        request.setAttribute("msg", msg);
        request.getRequestDispatcher("/admin/result.jsp").
        forward(request, response);
    }
}
```

【代码说明】

因为添加商品时，除了要添加商品字段外，还要同时上传商品图片，这时还要用到上传文件的包 commons-fileupload-1.3.3.jar。使用这个包时，首先定义一个 map 对象，用以接收表单传来的数据，同时要区分出是普通表单项，还是文件上传项。如果是文件上传项，还要判断上传文件的类型和大小，符合要求之后，还要指定唯一的文件。上传文件成功之后，还要把上传文件名和普通表单项数据封装到实体对象 shop 中，再调用商品业务逻辑方法 AddShop 来实现商品添加。

(五)商品添加子模块视图层设计

添加商品(addshop.jsp)页面设计。代码如下：

```jsp
<%@ taglib prefix="c" uri="http://java.sun.com/jsp/jstl/core" %>
<%@ page contentType="text/html;charset=UTF-8" language="java" %>
<!DOCTYPE html>
<html>
<head>
    <meta charset="UTF-8">
    <meta http-equiv="X-UA-Compatible" content="ie=edge">
    <title>添加商品</title>
    <link type="text/css" rel="stylesheet" href="style/reset.css">
    <link type="text/css" rel="stylesheet" href="style/addshop.css">
<script src="https://libs.baidu.com/jquery/2.1.4/jquery.min.js">
</script>
    <script type="text/javascript" src="js/addShop.js"></script>
</head>
<body>
<jsp:include page="head.jsp"/>
<div class="crumb">
    <div class="w">
        <div class="fl">
            <a href="index.jsp" class="link">QZMAIL</a>
            <span></span>
            <span class="link-text">${con}</span>
        </div>
        <div class="fr">
            <a href="exit.do" class="link">退出</a>
        </div>
    </div>
</div>
<div class="w">
    <jsp:include page="menu.jsp"/>
        <div class="content">
            <div class="content-con">
                <span class="text">商品添加 ${sucess}</span>
            </div>
    <form action="saveShop.do" onSubmit="javascript: return sendreg();" enctype="multipart/form-data" method="post" id="form">
```

```html
            <div class="shop-con">
                <ul>
                    <li><span class="label">商品品牌: </span><span class="con"><input id="shopname" type="text" name="shopname" /></span></li>
                    <li><span class="label">商品信息: </span><span class="con"><input id="shopinfo" type="text" name="shopinfo" style="width:200px;" /></span></li>
                    <li><span class="label">选择分类: </span> <span class="con">
                    <select name="sid" id="sid">
                        <c:forEach items="${categoryArrayList}" var="category">
                        <optgroup label="${category.cname}">
                            <c:forEach items="${category.subCates}" var="sname">
                            <c:if test="${category.cid==sname.category.cid}">
                            <option value="${sname.sid}">└─${sname.sname}</option>
                            </c:if>
                            </c:forEach>
                        </optgroup>
                        </c:forEach>
                    </select></span>
                    </li>
                    <li><span class="label">图 片 1:</span> <span class="con"><input id="image1" name="image1" type="file" /></span>
                    <span class="label">图 片 2:</span> <span class="con"><input id="image2" type="file" name="image2" /></span>
                    </li>
                    <li><span class="label">图 片 3:</span> <span class="con"><input id="image3" type="file" name="image3" /></span></li>
                    <li><span class="label">商品售价: </span> <span class="con"><input id="price" type="text" name="price" value="59.0" /></span> </li>
                    <li><span style="vertical-align: top;font-size:14px; ">商品详情: </span> <span class="con"><textarea rows="5" cols="40" wrap="hard" name="description" ></textarea> </span></li>
                </ul>
                <div class="add-btn"><input type="submit" id="btn" class="btn" value="提交"/></div>
            </div>
        </form>
    </div>
</div>
<jsp:include page="foot.jsp"/>
</body>
</html>
```

【代码说明】

(1) 添加商品时,由于要上传多张商品图片,而且要用到一个上传 jar 包,这个 jar 包要求我们必须把添加商品的信息分成普通表单项和上传文件表单项,商品信息提交时必须采用 post 方式提交。

(2) 添加商品信息时,要动态选择商品分类,本页面接收 AddShopServlet 传回的 categoryArrayList,运用 JSTL 标签进行解析,并应用到<select>选择框。

此页面运行效果如图 8-1 所示。

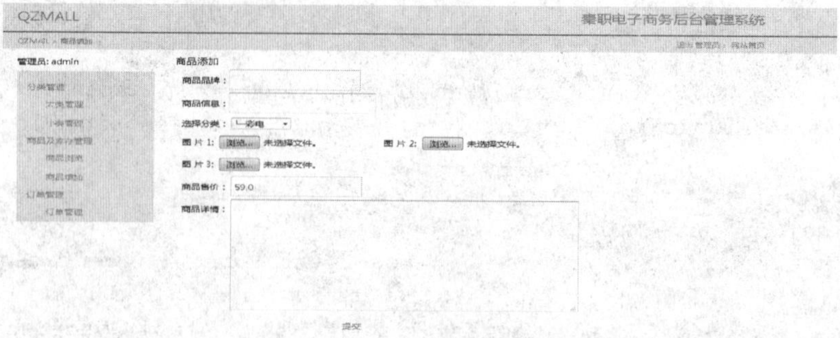

图 8-1　商品添加页面(addshop.jsp)

> **课堂技能训练：**
>
> 【实训操作内容】上传多张商品图片并改为唯一文件名称。
>
> 【实训操作要求】
>
> (1) 上传到指定目录 web/img 下。
>
> (2) 多文件上传，上传文件名称必须是唯一的。
>
> (3) 每个文件大小不能超过 500KB。

> 小提示：什么是 Spring Boot
>
> Spring Boot 是由 Pivotal 团队提供的全新框架，其设计目的是用来简化新 Spring 应用的初始搭建以及开发过程。该框架使用了特定的方式来进行配置，从而使开发人员不再需要定义样板化的配置。通过这种方式，Spring Boot 致力于在蓬勃发展的快速应用开发领域(rapid application development)成为领导者。
>
> (1) 创建独立的 Spring 应用程序。
>
> (2) 嵌入的 Tomcat，无须部署 WAR 文件。
>
> (3) 简化 Maven 配置。
>
> (4) 自动配置 Spring。
>
> (5) 提供生产就绪型功能，如指标、健康检查和外部配置。
>
> (6) 绝对没有代码生成和对 XML 没有要求配置。

第二节　商品维护子模块设计

一、商品维护基本流程

商品维护主要是指查找商品、更新商品和删除商品。

(1) 商品查询基本流程：管理员输入查询条件并提交查询；系统根据管理员提交的查询条件从数据库中查询商品并输出。

(2) 商品更新基本流程：管理员输入商品信息进行查询；管理员修改商品信息并提交；系统根据管理员提交的信息从数据库中更新商品信息。

(3) 商品删除基本流程：管理员输入商品的相关信息；系统根据管理员提交的信息从数据库中删除相关商品信息。

二、商品维护主要功能实现

(一)商品维护子模块持久层设计

1. 接口类设计

代码如下：

```java
public interface IShopDao {
    //查询所有商品及分页
    public List<Shop> getShopPage(int currentPage,int pageSize);
    public int getShopTotal();
    //更新单个产品
    public boolean updateByPid(Shop shop);  }
```

2. 接口实现类设计

商品接口实现类 ShopDaoImpl 位于 com.qzmall.dao.impl 包下，该类提供了关于表 shop 的查、删、改功能。代码如下：

```java
public class ShopDaoImpl implements IShopDao {
    /*
```

(1) 分页显示所有商品。

该方法用来查询表 shop 分页单位内商品记录功能，以请求页号和每页记录为参数，返回结果为商品集合分页边界类型。代码如下：

```java
    */
    @Override
 public List<Shop> getShopPage(int currentPage, int pageSize) {
        List<Shop> shops = new ArrayList<Shop>();
       String strSql = "select subcate.sname as sname, shop.* from shop,subcate  where shop.sid=subcate.sid order by shop.stock ASC ,shop.pid desc limit ?,?";
        return (List<Shop>) DBUtil.executeQuery(strSql, new IResultSetUtil() {
          @Override
    public Object doHandler(ResultSet rs) throws SQLException {
         while (rs.next()) {
            Shop shop = new Shop();
            SubCate subCate = new SubCate();
            shop.setPid(rs.getInt("pid"));
            subCate.setSname(rs.getString("sname"));
            shop.setSubCate(subCate);
```

```
            shop.setShopname(rs.getString("shopname"));
            shop.setShopinfo(rs.getString("shopinfo"));
            shop.setShopdate(rs.getString("shopdate"));
            shop.setStock(rs.getInt("stock"));
            shop.setPrice(rs.getFloat("price"));
            shop.setImage1(rs.getString("image1"));
            shops.add(shop);
            }
            return shops;
        }
    },(currentPage-1)*pageSize,pageSize);
}
/*
```

(2) 查询所有商品数目。

该方法用来查询表 shop 商品记录数的功能,返回结果为整数类型,表示记录数。代码如下:

```
*/
    @Override
public int getShopTotal() {
        String strSql = "select count(*) from shop ";
        Object obj= DBUtil.executeQuery(strSql);
        return Integer.parseInt(obj.toString());
}
/*
```

(3) 更新商品信息方法。

该方法用来实现更新商品表 shop 商品信息的功能,以商品实体对象为参数,返回结果为布尔类型,表示更新商品信息是否成功。代码如下:

```
*/
    @Override
    public boolean updateByPid(Shop shop) {
        String strSql = "update shop set
shopname=?,shopinfo=?,price=?,cid=?,sid=?,description=? where pid=?";
        return DBUtil.executeUpdate(strSql, shop.getShopname(),
shop.getShopinfo(), shop.getPrice(), shop.getCategory().getCid(),
shop.getSubCate().getSid(), shop.getDescription(), shop.getPid()) > 0;
    }
}
```

(二)商品维护子模块逻辑层设计

1. 接口类设计

接口类放在 com.qzmall.service.IShopService 类下,接口类设计的方法与持久层相同,这里不再赘述。

2. 接口实现类设计

接口类方法放在 com.qzmall.service.impl.ShopServiceImpl 类下。代码如下:

```java
public class ShopServiceImpl implements IShopService {
    @Override
    public Shop getShopByid(int pid) {
      IShopDao shopDao=new ShopDaoImpl();
      return shopDao.getShopByid(pid);}
    @Override
    public List<Shop> getShopPage(int currentPage, int pageSize)
{IShopDao shopDao=new ShopDaoImpl();
    return shopDao.getShopPage(currentPage,pageSize);}
    @Override
    public int getShopTotal() {
      IShopDao shopDao=new ShopDaoImpl();
      return shopDao.getShopTotal();}
    @Override
    public boolean updateByPid(Shop shop) {
      IShopDao shopDao=new ShopDaoImpl();
      return shopDao.updateByPid(shop); }
  }
```

(三)商品维护子模块控制层类设计

商品管理模块控制层类在 com.qzmall.servlet.shop 包下。

1. 商品列表功能设计

当管理员从 menu.jsp 页面发出商品维护的请求时，系统就会交给 BrowShopServlet 来处理，该类通过调用业务逻辑类的方法 getShopPage 来实现全部商品分页展示的功能。具体代码如下：

```java
@WebServlet(name = "BrowShopServlet",urlPatterns = "/admin/main.do")
public class BrowShopServlet extends HttpServlet {
    protected void doGet(HttpServletRequest req, HttpServletResponse resp)
throws ServletException, IOException {
        String con = "商品浏览";
        IShopService categoryService = new ShopServiceImpl();
        ArrayList<Category> categoryArrayList = categoryService.
getCategoryAll();
        req.setAttribute("categoryArrayList", categoryArrayList);
        try {
            int currentPage = Integer.parseInt(req.getParameter("currentPage"));
            System.out.println("CurrentPage="+currentPage);
            IShopService shopService = new ShopServiceImpl();
            int totalSize = shopService.getShopTotal();
            System.out.println("totalsize="+totalSize);
            int totalPage = totalSize / Tally.SHOP_PAGE_SIZE;
            if (totalSize % Tally.SHOP_PAGE_SIZE != 0)
                totalPage++;
            if (currentPage < 1) {
                currentPage = 1;
            }
            if (currentPage > totalPage) {
```

```
            currentPage = totalPage;
        }
    Pager page = new Pager(currentPage, totalSize);
    List<Shop> shopList = shopService.getShopPage(currentPage,
Tally.SHOP_PAGE_SIZE);
    req.setAttribute("con", con);
    req.setAttribute("shopList", shopList);
    req.setAttribute("page", page);
    String url = req.getRequestURI() + "?" req.getQueryString();
    System.out.println("url=" + url);
    int index = url.lastIndexOf("&currentPage=");
    if (index != -1) { url = url.substring(0,index);
        }
  System.out.println("urllast=" + url);
 req.setAttribute("url", url);
 req.getRequestDispatcher("main.jsp").forward(req, resp);
 }catch(NumberFormatException ex){
 String msg = "页码传输异常";
 req.setAttribute("msg", msg);
 req.getRequestDispatcher("result.jsp").forward(req, resp);
    }
  }
}
```

【代码说明】

在上述代码中，首先调用 categoryService.getCategoryAll() 得到所有分类对象 categoryArrayList，然后把它传回主页面 main.jsp，通过 req 获取显示商品页码 currentPage，通过调用 shopService.getShopTotal() 函数得到总记录数 totalSize，并根据全局变量 Tally.SHOP_PAGE_SIZE 得到总页数 totalPage，最后调用 Page 类 shopService.getShopPage 方法，得到 page 和 shopList，一并传回 main.jsp 页面。

2. 编辑浏览商品详细信息功能设计

当管理员通过 main.jsp 页面单击商品编号就会交给类 EditShopServlet 进行处理，该类通过调用商品业务逻辑类方法 getShopByid，来实现浏览商品详细信息功能。代码如下：

```
@WebServlet(name = "EditShopServlet",urlPatterns = "/admin/editShop.do")
public class EditShopServlet extends HttpServlet {
    protected void doGet(HttpServletRequest req, HttpServletResponse
resp) throws ServletException, IOException {
String con="商品编辑";
    req.setAttribute("con",con);
    int pid=Integer.parseInt(req.getParameter("pid"));
    IShopService shopService=new ShopServiceImpl();
    ArrayList<Category> categoryArrayList=shopService.getCategoryAll();
    Shop shop=shopService.getShopByid(pid);
    req.setAttribute("shop",shop);
    req.setAttribute("categoryArrayList",categoryArrayList);
    req.getRequestDispatcher("editshop.jsp").forward(req,resp);
    }
}
```

3. 更新商品功能设计

当管理员发出修改商品信息的请求后，就会交给类 UpdateShopServlet 进行处理，该类首先用 req 接收表单传过来的商品字段信息，然后封装这些信息于商品实体类 Shop 中，再通过调用商品业务逻辑类方法 updateByPid 来实现商品信息更新功能。代码如下：

```java
@WebServlet(name = "UpdateShopServlet",urlPatterns = "/admin/updateShop.do")
public class UpdateShopServlet extends HttpServlet {
    protected void doPost(HttpServletRequest req, HttpServletResponse resp) throws ServletException, IOException {
        ICategoryService categoryService=new CategoryServiceImpl();
        IShopService shopService=new ShopServiceImpl();
        int pid=Integer.parseInt(req.getParameter("pid"));
        String shopname=req.getParameter("shopname");
        String shopinfo=req.getParameter("shopinfo");
        int sid=Integer.parseInt(req.getParameter("sid"));
        SubCate subCate=new SubCate();
        subCate.setSid(sid);
        int cid=categoryService.getCid(sid);
        Category category=new Category();
        category.setCid(cid);
        float price=Float.parseFloat(req.getParameter("price"));
        String description=req.getParameter("description");
        Shop shop=new Shop();
        shop.setPid(pid);
        shop.setShopname(shopname);
        shop.setShopinfo(shopinfo);
        shop.setPrice(price);
        shop.setCategory(category);
        shop.setSubCate(subCate);
        shop.setDescription(description);
        if (shopService.updateByPid(shop)){
            String msg="更新商品信息成功！";
            req.setAttribute("msg",msg);
    req.getRequestDispatcher("result.jsp").forward(req,resp);
        }else{
            String msg="更新商品信息失败！";
            req.setAttribute("msg",msg);
    req.getRequestDispatcher("result.jsp").forward(req,resp);
        }
    }
}
```

(四)商品维护子模块视图层设计

1. 商品列表显示主页面(main.jsp)设计

代码如下：

```jsp
<%@ taglib prefix="c" uri="http://java.sun.com/jsp/jstl/core" %>
<%@ page contentType="text/html;charset=UTF-8" language="java" %>
<!DOCTYPE html>
<html>
<head>
    <meta charset="UTF-8">
    <meta http-equiv="X-UA-Compatible" content="ie=edge">
    <title>产品编辑</title>
    <link type="text/css" rel="stylesheet" href="style/reset.css">
    <link type="text/css" rel="stylesheet" href="style/main.css">
    <script src="https://libs.baidu.com/jquery/2.1.4/jquery.min.js"></script>
    <script type="text/javascript">
        function querysubcate(){
            var sid = $("#sid").val();
            location.href='adminQuerySubShop.do?sid='+sid+'&currentPage=1';
        }
        $(function(){
            $(".jian").click(function(){
                var dd= $(this).attr("id").indexOf("jian")
                var pid=$(this).attr("id").substring(0,dd);
                var quantity=$("#"+pid+"num").val();
                if(quantity==1){
                    alert("采购数量不能低于1")
                    return false;
                }
                $("#" + pid + "num").val(Number(quantity)-1);
            });
            $(".jia").click(function(){
                var dd= $(this).attr("id").indexOf("jia")
                var pid=$(this).attr("id").substring(0,dd);
                var quantity=$("#"+pid+"num").val();
                if (Number(quantity)>50){
                    alert("采购数量大于库存!")
                    return false;
                }
                $("#" + pid + "num").val(Number(quantity)+1);
            });
        })
    </script>
</head>
<body>
<jsp:include page="head.jsp"/>
  <div class="crumb">
    <div class="w">
        <div class="fl">
            <a href="index.jsp" class="link">QZMAIL</a>
            <span>></span>
            <span class="link-text">${con}</span>
        </div>
```

```html
            <div class="fr">
                <a href="exit.do" class="link">退出</a>
            </div>
        </div>
</div>
    <div class="w">
        <jsp:include page="menu.jsp"/>
        <div class="content">
            <div class="content-con">
                <form action="adminQueryShop.do" method="get" >
                <input type="text" value="" name="shopname" class="search">
                    <input class="currentPage" id="currentPage" name="currentPage" value="1" type="hidden"/>
                    <input type="submit" class="btn search-btn" id="search-btn" value="搜索"></input>
                </form>
    <span class="text"><a class="link" href="main.do?&currentPage=1">全部商品</a></span>
<span class="text">商品分类: </span><select name="sid" id="sid" class="select" onchange="querysubcate();">
        <option value="" selected="selected">==请选择2级分类====</option>
        <c:forEach items="${categoryArrayList}" var="category">
        <optgroup label="${category.cname}">
        <c:forEach items="${category.subCates}" var="sname">
        <c:if est="${category.cid==sname.category.cid}">
        <option value="${sname.sid}">└${sname.sname}</option>
                    </c:if>
                </c:forEach>
            </optgroup>
        </c:forEach>
    </select>
    </div>
     <form action="addStockItem.do" method="post">
    <table class="table" cellspacing="0" cellpadding="0">
        <thead>
        <tr>
            <th width="10%">选择</th>
            <th width="15%">商品名称</th>
            <th width="15%">商品图片</th>
            <th width="15%">商品类别</th>
            <th width="10%">市场价</th>
            <th width="10%">库存</th>
            <th width="25%">操作</th>
        </tr>
        </thead>
        <tbody>
        <c:forEach items="${shopList}" var="shop">
            <tr>
            <td width="80px"><input value="${shop.pid}" name="checkboxBtn" type="checkbox"/></td>
```

```
                <td width="120px"><a class="link" href= "editShop.do?pid=
${shop.pid}">${shop.shopname}</a></td>
                <td width="120px"><img src="<c:url
value='../img/pic/${shop.image1}'/>" width="80" height="80" ></td>
                <td width="120px">${shop.subCate.sname}</td>
                <td width="80px">${shop.price}</td>
                <td width="80px">${shop.stock}</td>
                <td width="200px"><a class="link" href="deleShop.do?pid=
${shop.pid}" onClick="return confirm('您确定进行删除操作吗？')">删除</a>
                <a class="link" href="stockItemByPid.do?pid=
${shop.pid}&currentPage=1">库存明细</a></td>
              </tr>
            </c:forEach>
          </tbody>
        </table>
        <div class="stock"><input type="submit" class="btn" value="增加库
存"></div>
      </form>
      <div class="pg-content">
          <form class="page-form" method="post" action="${url}" >
            <input type="text" name="currentPage" class="page" >
            <input type="submit" value="go" >
          </form>
 <c:if test="${page.hasFirst}">
          <li><a href="${url}&currentPage=1">首页</a></li>
 </c:if>
 <c:if test="${page.hasPrevious}">
 <li><a href="${url}&currentPage=${page.currentPage-1}">上一页</a></li>
 </c:if>
 <c:if test="${page.hasNext}">
 <li><a href="${url}&currentPage=${page.currentPage+1}">下一页/a></li>
 </c:if>
 <c:if test="${page.hasLast}">
 <li><a href="${url}&currentPage=${page.totalPage}">尾页</a></li>
 </c:if>
 <span class="pg-total">当前第${page.currentPage}页，总共
${page.totalPage}页 </span>
          </div>
        <!--分页结束-->
    </div>
 </div>
<jsp:include page="foot.jsp"/>
</body>
</html>
```

【代码说明】

此页面是后台管理的主页面，主要包括以下几个方面的功能。

(1) 显示全部商品。通过 main.do 命令调用 BrowShopServlet 类，用 JSTL 标签遍历返回对象 shoplist.page 来完成此功能。

(2) 模糊查询商品。通过 adminQueryShop.do 调用 AdminQueryShopServlet 类，用 JSTL 标签遍历返回对象 shoplist.page 来完成此功能。

(3) 查询某一子类商品。通过 jQuery 的方法 querysubcate()调用 AdminQuerySubShopServlet 类，用 JSTL 标签遍历返回对象 shoplist.page 来完成此功能。运行效果如图 8-2 所示。

图 8-2　后台主页面(main.jsp)

2. 商品编辑页面(editshop.jsp)设计

代码如下：

```jsp
<%@ taglib prefix="c" uri="http://java.sun.com/jsp/jstl/core" %>
<%@ page contentType="text/html;charset=UTF-8" language="java" %>
<!DOCTYPE html>
<html>
<head>
    <meta charset="UTF-8">
    <meta http-equiv="X-UA-Compatible" content="ie=edge">
    <title>商品编辑</title>
    <link type="text/css" rel="stylesheet" href="style/reset.css">
    <link type="text/css" rel="stylesheet" href="style/addshop.css">
<script src="https://libs.baidu.com/jquery/2.1.4/jquery.min.js"></script>
    <script type="text/javascript" src="js/addShop.js"></script>
</head>
<body>
<jsp:include page="head.jsp"/>
<div class="crumb">
    <div class="w">
        <div class="fl">
            <a href="index.jsp" class="link">QZMAIL</a>
            <span>></span>
            <span class="link-text">${con}</span>
        </div>
```

```html
             <div class="fr">
                 <a href="exit.do" class="link">退出</a>
             </div>
         </div>
</div>
<div class="w">
     <jsp:include page="menu.jsp"/>
     <div class="content">
         <div class="content-con">
             <span class="text">编辑商品</span>
         </div>
         <form action="updateShop.do?pid=${shop.pid}" onSubmit="javascript:return sendreg();" method="post" id="form">
             <div class="shop-con">
                 <ul>
                     <li><span class="label">商品品牌：</span><span class="con"><input id="shopname" type="text" name="shopname" value="${shop.shopname}" /></span></li>
                     <li><span class="label">商品信息：</span><span class="con"><input id="shopinfo" type="text" name="shopinfo" value="${shop.shopinfo}" style="width:200px;" /></span></li>
                     <li><span class="label">选择分类：</span> <span class="con">
                         <select name="sid" id="sid">
                     <c:forEach items="${categoryArrayList}" var="category">
                         <optgroup label="${category.cname}">
                         <c:forEach items="${category.subCates}" var="sname">
                          <c:if test="${category.cid==sname.category.cid}">
                            <c:if test="${shop.subCate.sid==sname.sid}">
                              <option value="${sname.sid}" selected="selected">${sname.sname}</option>
                            </c:if>
                            <option value="${sname.sid}">${sname.sname}</option>
                          </c:if>
                         </c:forEach>
                         </optgroup>
                     </c:forEach>
                         </select></span>
                     </li>
                     <li>
                       <span class="label">商品售价：</span> <span class="con"><input id="price" type="text" name="price" value="${shop.price}" /></span>
                     </li>
                     <li><span style="vertical-align: top;font-size:14px; ">商品详情：</span> <span class="con"><textarea rows="5" cols="40" wrap="hard" name="description" >${shop.description}</textarea> </span></li>
                 </ul>
```

```
            <div class="add-btn"><input type="submit" id="btn" class="btn" value="提交"/></div>
          </div>
      </form>
    </div>
</div>
<jsp:include page="foot.jsp"/>
</body>
</html>
```

【代码说明】

此页面是根据商品 id 查找某一商品的详细信息，通过 EditShopServlet 返回 shop 对象来遍历商品信息，通过 EditShopServlet 返回的 categoryArrayList 来遍历商品分类信息。另外，根据这个页面数据对商品信息进行编辑修改，此页面的难点在于能够正确显示所选商品的分类，在此基础上，还能动态选择商品分类并进行商品信息修改。商品编辑页面运行效果如图 8-3 所示。

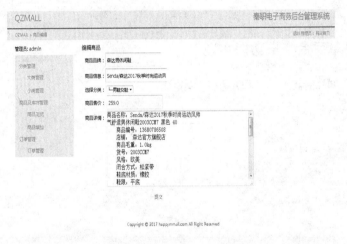

图 8-3　商品编辑页面(editshop.jsp)

课堂技能训练：

【实训操作内容】完成后台的商品更新功能。

【实训操作步骤】

(1) 持久层设计(重点)。

(2) 业务逻辑层设计。

(3) 控制层设计。

(4) 视图层设计。

小提示：Eclipse 平台

Eclipse 是一个开放源代码的、基于 Java 的可扩展开发平台。Eclipse 既包括作为 Java 集成开发环境(IDE)来使用，还包括插件开发环境(Plug-in Development Environment，

PDE)，这个组件主要针对希望扩展 Eclipse 的软件开发人员，因为它允许用户构建与 Eclipse 环境无缝集成的工具。由于 Eclipse 中的每样东西都是插件，对于给 Eclipse 提供插件，以及给用户提供一致和统一的集成开发环境而言，所有工具开发人员都具有同等的发挥场所。

第三节 用户订单后台维护子模块设计

一、用户订单维护基本流程

订单浏览基本流程：管理员登录；浏览所有已付款订单。

订单查询基本流程：管理员输入查询条件并提交查询；系统根据管理员提交的查询条件从数据库中查出相关订单信息。

订单状态修改基本流程：管理员输入查询条件；从中选择某一订单；进入订单明细；修改订单状态并提交。

二、用户订单后台维护功能实现

(一)用户订单后台维护子模块持久层设计

1. 接口类设计

代码如下：

```java
public interface IOrdersDao {
    //已付款订单
    public List<Orders> allOrders(int currentPage, int pageSize);
    public int getAllOrdersTotal();
}
```

2. 接口实现类设计

代码如下：

```java
public class OrderDaoImpl implements IOrdersDao {
/*
```

(1) 该方法用来实现分页查询订单表 orders 订单状态大于 1 的所有订单，以请求页码和每页记录数为参数，返回结果为分页边界类型。具体代码如下：

```java
*/
@Override
public List<Orders> allOrders(int currentPage, int pageSize)
{ List<Orders> orders = new ArrayList<Orders>();
    String strSql = "select * from orders where state>1 order by state ASC limit ?,?";
    return (List<Orders>) DBUtil.executeQuery(strSql, new IResultSetUtil()
{
```

```
    @Override
      public Object doHandler(ResultSet rs) throws SQLException {
            while (rs.next()) {
                Orders orders1 = new Orders();
orders1.setUsername(rs.getString("username"));
orders1.setTruename(rs.getString("truename"));
orders1.setOrderId(rs.getString("orderId"));
                orders1.setPhone(rs.getString("phone"));
                orders1.setSum(rs.getFloat("sum"));
                orders1.setState(rs.getInt("state"));
                orders1.setAddtime(rs.getString("addtime"));
                orders.add(orders1);
            }
            return orders;
        }
    }, (currentPage - 1) * pageSize, pageSize);
   }
@Override
/*
```

(2) 该方法用来实现分页查询订单表 orders 订单状态大于 1 的所有订单数量，返回结果为整数类型。具体代码如下：

```
*/
   public int getAllOrdersTotal() {
   String strSql = "select count(*) from orders where state>1 ";
        Object obj= DBUtil.executeQuery(strSql);
        return Integer.parseInt(obj.toString());
    }
}
```

(二)用户订单后台维护子模块逻辑层设计

1. 接口类设计

接口类放在 com.qzmall.service.IOrderService 类下，接口类设计的方法与持久层相同，这里不再赘述。

2. 接口实现类设计

代码如下：

```
public class OrdersServiceImpl implements IOrdersService {
    @Override
    public List<Orders> allOrders(int currentPage, int pageSize)
{IOrdersDao ordersDao=new OrderDaoImpl();
     return ordersDao.allOrders(currentPage, pageSize); }
    @Override
    public int getAllOrdersTotal() {
    IOrdersDao ordersDao=new OrderDaoImpl();
    return ordersDao.getAllOrdersTotal();}
}
```

(三)用户订单后台维护控制层设计

1. 查询所有已付款订单

当管理员在 left.jsp 页面发出订单管理的请求后,就会交给 AdminAllOrdersServlet 来进行处理,该类通过调用 Page 类,以及商品业务逻辑类的 getAllOrdersTotal、allOrders 方法来实现查询所有已付款的订单的功能。代码如下:

```java
@WebServlet(name = "AdminAllOrdersServlet",urlPatterns =
"/admin/adminAllOrders.do")
public class AdminAllOrdersServlet extends HttpServlet {
    protected void doPost(HttpServletRequest req, HttpServletResponse resp) throws ServletException, IOException {
        doGet(req,resp);}
    protected void doGet(HttpServletRequest req, HttpServletResponse resp) throws ServletException, IOException {
        String con = "已付款订单";
        try {
            int currentPage = Integer.parseInt(req.getParameter("currentPage"));
            System.out.println("CurrentPage="+currentPage);
            IOrdersService ordersService = new OrdersServiceImpl();
            int totalSize = ordersService.getAllOrdersTotal();
            System.out.println("totalsize="+totalSize);
            int totalPage = totalSize / Tally.SHOP_PAGE_SIZE;
            if (totalSize % Tally.SHOP_PAGE_SIZE != 0)
                totalPage++;
            if (currentPage < 1) {
                currentPage = 1;
            }
            if (currentPage > totalPage) {
                currentPage = totalPage;
            }
            Pager page = new Pager(currentPage, totalSize);
            List<Orders> ordersInfo = ordersService.allOrders(currentPage, Tally.SHOP_PAGE_SIZE);
            req.setAttribute("con", con);
            req.setAttribute("ordersInfo", ordersInfo);
            req.setAttribute("page", page);
            String url = req.getRequestURI() + "?" + req.getQueryString();
            System.out.println("url=" + url);
            int index = url.lastIndexOf("&currentPage=");
            if (index != -1) {
                url = url.substring(0, index);
            }
            System.out.println("urllast=" + url);
            req.setAttribute("url", url);
            req.getRequestDispatcher("adminOrders.jsp").forward(req, resp);
        }catch(NumberFormatException ex){
            String msg = "页码传输异常";
```

```
            req.setAttribute("msg", msg);
            req.getRequestDispatcher("result.jsp").forward(req, resp);
        }
    }
}
```

2. 根据订单编号查询订单功能设计

当管理员在 adminOrders.jsp 页面发出查询订单编号的请求后，就会交给 AdminOneOrdersServlet 来进行处理，该类通过调用 getOrdersByOrderId 返回查询结果 orders，然后把它传回 adminOneOrders.jsp 页面。代码如下：

```
@WebServlet(name = "AdminOneOrdersServlet",urlPatterns = "/admin/adminQueryOrderId.do")
public class AdminOneOrdersServlet extends HttpServlet {
    protected void doPost(HttpServletRequest req, HttpServletResponse resp) throws ServletException, IOException { doGet(req,resp); }
    protected void doGet(HttpServletRequest req, HttpServletResponse resp) throws ServletException, IOException {
        String orderId=req.getParameter("orderId");
        if (orderId!="" && orderId!=null){
            IOrdersService ordersService=new OrdersServiceImpl();
            Orders orders=ordersService.getOrdersByOrderId(orderId);
            req.setAttribute("orders",orders);
            req.getRequestDispatcher("adminOneOrders.jsp").forward(req,resp);
        }
        else{
            String msg="订单编号不能为空！";
            req.setAttribute("msg",msg);
            req.getRequestDispatcher("result.jsp").forward(req,resp);
        }
    }
}
```

3. 根据订单编号查询订单明细功能设计

当管理员在 adminOrders.jsp 页面发出订单明细的请求后，就会交给 AdminOrdersListServlet 来进行处理，该类通过调用 getOrdersByOrderId 返回查询结果 orders，然后把它传回 adminOrdersList.jsp 页面。代码如下：

```
@WebServlet(name = "AdminOrdersListServlet",urlPatterns = "/admin/adminOrdersList.do")
public class AdminOrdersListServlet extends HttpServlet {
    protected void doPost(HttpServletRequest req, HttpServletResponse resp) throws ServletException, IOException { doGet(req,resp);}
    protected void doGet(HttpServletRequest req, HttpServletResponse resp) throws ServletException, IOException {
        String orderId=req.getParameter("orderId");
        IOrdersService ordersService=new OrdersServiceImpl();
        Orders orders=ordersService.getOrdersByOrderId(orderId);
```

```
        List<OrdersItem> ordersItemList=ordersService.getOrdersItemByOrderId(orderId);
        req.setAttribute("orders",orders);
        req.setAttribute("ordersItemList",ordersItemList);
        req.getRequestDispatcher("adminOrdersList.jsp").forward(req,resp);
    }
}
```

4．更新订单状态功能设计

当管理员在 adminOrdersList.jsp 页面发出更新订单状态的请求后，就会交给 AdminUpdateOrderStateServlet 来进行处理，实现订单状态的更新。代码如下：

```
@WebServlet(name = "AdminUpdateOrderStateServlet",urlPatterns = "/admin/adminUpdateOrderState.do")
public class AdminUpdateOrderStateServlet extends HttpServlet {
    protected void doPost(HttpServletRequest req, HttpServletResponse resp) throws ServletException, IOException {
        String orderId = req.getParameter("orderId");
        if (states==null ||"".equals(states)){
            String msg = "订单状态未做任何改变";
            req.setAttribute("msg", msg);
            req.getRequestDispatcher("result.jsp").forward(req, resp);
        }
        int state=Integer.parseInt(states);
        Orders orders = new Orders();
        orders.setOrderId(orderId);
        orders.setState(state);
        IOrdersService ordersService = new OrdersServiceImpl();
        if (ordersService.updateOrdersState(orders)) {
            IShopService shopService = new ShopServiceImpl();
            List<OrdersItem> ordersItemList = new ArrayList<OrdersItem>();
            ordersItemList = ordersService.getOrdersItemByOrderId(orderId);
            Iterator iter = ordersItemList.iterator();
            while (iter.hasNext()) {
                OrdersItem ordersItem=(OrdersItem)iter.next();
                int shopnum = ordersItem.getShopnum();
                int pid=ordersItem.getPid();
                StockItem stockItem = new StockItem();
                stockItem.setPid(pid);
                stockItem.setNum(-shopnum);
                stockItem.setStockTime(StringHelper.getCurrentFormatDate());
                stockItem.setDescription("卖出商品");
                shopService.addstock(stockItem);
            }
            String msg = "修改订单状态成功,并相应减少库存";
            req.setAttribute("msg", msg);
            req.getRequestDispatcher("result.jsp").forward(req, resp);
        }
        else{
            String msg="修改订单状态未成功";
```

```
            req.setAttribute("msg",msg);
req.getRequestDispatcher("result.jsp").forward(req,resp);
    }
  }
protected void doGet(HttpServletRequest req, HttpServletResponse resp)
throws ServletException, IOException {  doPost(req,resp);}
}
```

【代码说明】

(1) 运用 req 对象接收订单状态和订单编号,然后封装到订单实体类对象(orders)中。

(2) 管理员修改状态就一种,由订单状态 2(已付款订单)改为订单状态 3(已发货)。

(3) 当管理员发出改变订单状态的请求时,调用业务逻辑类 updateOrdersState(orders) 方法,进行订单状态更新,如果返回值为真,此时的订单状态更新为"已发货",所以还要从库存表相应减少订单中所包含商品的库存数量。这时还要调用商品业务逻辑类方法 shopService.addstock(stockItem),在库存表中增加商品库存记录,因为库存减少意味着现库存量减去原库存量所得值为负数,所以在封装库存实体类时代码应写成: stockItem.setNum (-shopnum)。

(四)用户订单后台维护子模块视图层设计

1. 显示所有已付款订单页面(adminOrders.jsp)设计

代码如下:

```
<%@ taglib prefix="c" uri="http://java.sun.com/jsp/jstl/core" %>
<%@ page contentType="text/html;charset=UTF-8" language="java" %>
<!DOCTYPE html>
<html>
<head>
    <meta charset="UTF-8">
    <meta http-equiv="X-UA-Compatible" content="ie=edge">
    <title>订单编辑</title>
    <link type="text/css" rel="stylesheet" href="style/reset.css">
    <link type="text/css" rel="stylesheet" href="style/main.css">
    <script src="https://libs.baidu.com/jquery/2.1.4/jquery.min.js"></script>
</head>
<body>
<jsp:include page="head.jsp"/>
<div class="crumb">
    <div class="w">
        <div class="fl">
            <a href="index.jsp" class="link">QZMAIL</a>
            <span></span>
            <span class="link-text">${con}</span>
        </div>
        <div class="fr">
            <a href="exit.do" class="link">退出</a>
```

```jsp
            </div>
        </div>
</div>
<div class="w">
    <jsp:include page="menu.jsp"/>
    <div class="content">
        <div class="content-con">
            <form action="adminQueryOrderId.do" method="get" >
                <label class="label">订单编号: </label> <input type="text" value="" name="orderId" class="search">
                <input type="submit" class="btn search-btn" id="search-btn" value="搜索"></input>
            </form>
            <span class="text"><a href="adminAllOrders.do?currentPage=1" class="link">全部订单</a></span>
        </div>
         <table class="table" cellspacing="0" cellpadding="0">
            <thead>
            <tr>
                <th width="140px">订单编号</th>
                <th width="120px">真实姓名</th>
                <th width="100px">手机</th>
                <th width="80px">金额</th>
                <th width="160px">时间</th>
                <th width="80px">状态</th>
                <th width="120px">操作</th>
            </tr>
            </thead>
            <tbody>
            <c:forEach items="${ordersInfo}" var="orders">
                <tr>
                    <td width="140px">${orders.orderId}</td>
                    <td width="120px">${orders.truename}</td>
                    <td width="100px">${orders.phone}</td>
                    <td width="80px">${orders.sum}</td>
                    <td width="160px">${orders.addtime}</td>
                    <td width="80px">
                       <c:if test="${orders.state==1}">未付款</c:if>
                       <c:if test="${orders.state==2}">已付款</c:if>
                       <c:if test="${orders.state==3}">已发货</c:if>
                       <c:if test="${orders.state==4}">订单完成</c:if>
                    </td>
                    <td width="120px"><a class="link" href="adminOrdersList.do?orderId=${orders.orderId}">订单明细</a></td>
                </tr>
            </c:forEach>
            </tbody>
         </table>
        <div class="pg-content">
            <form class="page-form" method="post" action="${url}" >
```

```
                <input type="text" name="currentPage" class="page" >
                <input type="submit" value="go" >
            </form>
            <c:if test="${page.hasFirst}">
                <li><a href="${url}&currentPage=1">首页</a></li>
            </c:if>
            <c:if test="${page.hasPrevious}">
                <li><a href="${url}&currentPage=
                ${page.currentPage-1}">上一页</a></li>
            </c:if>
            <c:if test="${page.hasNext}">
                <li><a href="${url}&currentPage=
                 ${page.currentPage+1}">下一页</a></li>
            </c:if>
            <c:if test="${page.hasLast}">
                <li><a href="${url}&currentPage=
                    ${page.totalPage}">尾页</a></li>
            </c:if>
            <span class="pg-total">当前第${page.currentPage}页，总共
${page.totalPage}页 </span>
        </div>
        <!--分页结束-->
    </div>
</div>
<jsp:include page="foot.jsp"/>
</body>
</html>
```

【代码说明】

该页面显示的是所有已付款订单，并进行分页显示，该页面接收 AdminAllOrdersServlet 传回的分页集合对象 ordersInfo、分页容器对象 page 及地址栏变量 url，通过 JSTL 标签进行遍历解析，把内容显示到当前页面上，运行效果如图 8-4 所示。

图 8-4　所有已付款订单

2. 查询单个订单页面(adminOneOrders.jsp)设计

代码如下：

```jsp
<%@ taglib prefix="c" uri="http://java.sun.com/jsp/jstl/core" %>
<%@ page contentType="text/html;charset=UTF-8" language="java" %>
<!DOCTYPE html>
<html>
<head>
    <meta charset="UTF-8">
    <meta http-equiv="X-UA-Compatible" content="ie=edge">
    <title>订单维护</title>
    <link type="text/css" rel="stylesheet" href="style/reset.css">
    <link type="text/css" rel="stylesheet" href="style/main.css">
    <script src="https://libs.baidu.com/jquery/2.1.4/jquery.min.js"></script>
</head>
<body>
<jsp:include page="head.jsp"/>
<div class="crumb">
    <div class="w">
        <div class="fl">
            <a href="index.jsp" class="link">QZMAIL</a>
            <span>></span>
            <span class="link-text">${con}</span>
        </div>
        <div class="fr">
            <a href="exit.do" class="link">退出</a>
        </div>
    </div>
</div>
<div class="w">
    <jsp:include page="menu.jsp"/>
    <div class="content">
        <div class="content-con">
            <form action="adminQueryOrderId.do" method="get" >
            <label class="label">订单编号: </label> <input type="text" value="" name="shopname" class="search">
            <input type="submit" class="btn search-btn" id="search-btn" value="搜索"></input>
            </form>
            <span class="text"><a href="adminAllOrders.do?currentPage=1">全部订单</a></span>
        </div>
        <table class="table" cellspacing="0" cellpadding="0">
            <thead>
            <tr>
                <th width="15%">订单编号</th>
                <th width="15%">真实姓名</th>
                <th width="10%">手机</th>
                <th width="10%">金额</th>
                <th width="15%">时间</th>
                <th width="10%">状态</th>
                <th width="25%">操作</th>
            </tr>
            </thead>
            <tbody>
```

```
                    <tr>
                        <td width="15%">${orders.orderId}</td>
                        <td width="15%">${orders.truename}</td>
                        <td width="15%">${orders.phone}</td>
                        <td width="10%">${orders.sum}</td>
                        <td width="15%">${orders.addtime}</td>
                        <td width="10%"><c:if test="${orders.state==1}">未付款
</c:if>
            <c:if test="${orders.state==2}">已付款</c:if>
            <c:if test="${orders.state==3}">已发货</c:if>
            <c:if test="${orders.state==4}">订单完成</c:if>
                        </td>
                        <td width="20%"><a class="link"
href="adminOrdersList.do?orderId=${orders.orderId}">订单明细</a></td>
                    </tr>
                </tbody>
            </table>
        </div>
</div>
<jsp:include page="foot.jsp"/>
</body>
</html>
```

【代码说明】

该页面显示的是某一订单编号的列表信息，该页面接收 AdminOneOrdersServlet 传回的订单主表实体类对象 orders，通过 EL 标签进行遍历解析，把内容显示到当前页面上，运行效果如图 8-5 所示。

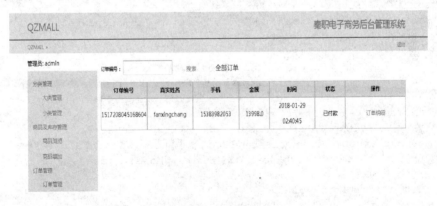

图 8-5 查询某一订单编号结果

3. 订单明细页面(adminOrdersList.jsp)设计

代码如下：

```
<%@ taglib prefix="c" uri="http://java.sun.com/jsp/jstl/core" %>
<%@ page contentType="text/html;charset=UTF-8" language="java" %>
<!DOCTYPE html>
<html>
```

```html
<head>
    <meta charset="UTF-8">
    <meta http-equiv="X-UA-Compatible" content="ie=edge">
    <title>产品编辑</title>
    <link type="text/css" rel="stylesheet" href="style/reset.css">
    <link type="text/css" rel="stylesheet" href="style/main.css">
    <script src="https://libs.baidu.com/jquery/2.1.4/jquery.min.js"></script>
</head>
<body>
<jsp:include page="head.jsp"/>
<div class="crumb">
    <div class="w">
        <div class="fl">
            <a href="index.jsp" class="link">QZMAIL</a>
            <span>></span>
            <span class="link-text">${con}</span>
        </div>
        <div class="fr">
            <a href="exit.do" class="link">退出</a>
        </div>
    </div>
</div>
<div class="w">
    <jsp:include page="menu.jsp"/>
    <div class="content">
        <div class="panel">
        <div class="panel-title">修改订单状态</div> <span id="span12">${resu}</span>
        <form name="form1" method="post" action="adminUpdateOrderState.do" >
            <div class="panel-body">
                <div class="form-line">
                    <span class="label">收货人：</span>
                    <span class="text">${orders.truename}</span>
        <input id="orderId" name="orderId" value="${orders.orderId}" type="hidden" readOnly="true"/>
                                <span class="label">手  机：</span>
                    <span class="text">${orders.phone}</span>
                </div>
    <div class="form-line">
        <span class="label">地  址：</span>
        <span class="text">${orders.address}</span> 
        <span class="label">邮  编：</span>
        <span class="text">${orders.postcode}</span>
    </div>
     <div class="form-line">
        <span class="label">金  额：</span>
        <span class="text"><input type="text" id="sum" name="sum" value="${orders.sum}" readOnly="true"></span>
        <span class="label">状  态：</span>
        <span class="text"><select name="state">
        <c:if test="${orders.state==1}">
```

```html
<option value="1" selected="selected"  disabled="disabled">未付款</option>
<option value="2" selected="selected"  disabled="disabled">已付款</option>
            </c:if>
            <c:if test="${orders.state==2}">
<option value="2" selected="selected"  disabled="disabled">客户已付款</option>
            <option value="3" >商家已发货</option>
             </c:if>
             <c:if test="${orders.state==3}">
<option value="3" selected="selected"  disabled="disabled">商家已发货</option>
            <option value="4" disabled="disabled">用户已收货</option>
             </c:if>
             <c:if test="${orders.state==4}">
<option value="4" selected="selected"  disabled="disabled">用户已收货</option>
            </c:if>
            </select></span>
            </div>
            <div class="form-line"> <span class="text"><input class="btn btn-con" type="submit" value="提交"></span>
            </div>
           </div>
           <div class="panel-body">
            <table class="table" cellspacing="0" cellpadding="0">
             <thead>
              <tr>
               <th width="20%">订单编号</th>
               <th width="20%">商品编号</th>
               <th width="20%">商品名称</th>
               <th width="20%">数量</th>
               <th width="20%">价格</th>
              </tr>
             </thead>
             <tbody>
              <c:forEach items="${ordersItemList}" var="item">
               <tr>
                <td width="20%">${item.orderId}</td>
<td width="20%"><input name="pid" value="${item.pid}" readonly="true"/></td>
<td width="20%"><input name="shopname" value="${item.shopname}" readonly="true"/> </td>
<td width="20%"><input name="shopnum" value="${item.shopnum}" readonly="true"/> </td>
                <td width="20%">${item.price}</td>
               </tr>
              </c:forEach>
             </tbody>
            </table>
           </div>
          </form>
        </div>
    </div>
```

```
</div>
<jsp:include page="foot.jsp"/>
</body>
</html>
```

【代码说明】

该页面接收 AdminOrdersListServlet 传回的订单主表实体类对象 order 和订单明细集合类 ordersItemList，通过 EL 和 JSTL 标签进行遍历解析，把某一订单编号的内容显示到当前页面上，运行效果如图 8-6 所示。

图 8-6　订单明细页面

当我们改变订单状态，更改为"商家已发货"时，单击"提交"按钮，出现如图 8-7 所示的页面，表示订单状态修改成功。

图 8-7　修改订单状态成功

课堂技能训练：

【实训操作内容】完成后台订单维护功能。

【实训操作步骤】

(1) 持久层设计。

(2) 业务逻辑层设计。

(3) 控制层设计。

(4) 视图层设计。

要求：能够显示所有已付款订单列表，查询订单，以及修改订单状态。

课后训练

一、任务要求

（1）qzmall 电子商城后台管理目录为 admin，做一个过滤器，使没有登录的用户无法访问这个目录。

（2）使用 JSP+Servlet+JavaBean 构成的 MVC 模型，完成手机管理系统。

二、任务描述

(一)语言和环境

实现技术：Java Web 技术。

环境要求及开发工具：JDK 1.7 以上、Eclipse 或 IntelliJ IDEA. Tomcat 8.0 以上。

(二)程序整体要求

主要功能如下。

（1）手机添加。手机包括手机 ID、手机名、操作系统、手机图片、价格、手机描述，要将手机信息保存到集合中，同时手机图片上传到服务器。

（2）手机信息查询。包括显示所有手机信息和根据手机名查询手机信息并显示。

（3）手机修改。根据手机 id 进行手机修改，如果不存在该手机则在页面中显示 id 不存在的提示。

（4）手机删除。根据手机 id 进行手机删除，如果不存在该手机则在页面中显示 id 不存在的提示。

注意：数据存储到集合中(使用 ArrayList、HashSet 和 HashMap 集合均可)。

(三)思路分析

由场景和运行效果，可以分析出项目中可以抽取如下类。

1. 手机类 Mob

类型描述：能够描述手机 ID、手机名、操作系统、图片地址、价格等。

方法：构造方法、get 和 set 方法、toString()方法。

2. 处理上传相关的类 Uploadutils

类型描述：存储添加页面中所有参数的集合 Map<String, Object> params;，其中 key 值是表单中控件的名称，value 是表单控件对应的值。

方法：文件上传的工具方法：public static Map<String, Object> UploadFile(HttpServletRequest request，String uploadDirectory)。

提示：该方法负责将表单中的信息存储到集合中，并完成图片的上传。

3. 手机数据处理类 MobDaoImpl

类型描述：存放手机信息的 List：private static final List<Map<String, Object>> db。

提示：Map 中存储的就是一条手机的信息，也就是 Uploadutils 中的 Map。

方法：添加手机：public void addMob(Map<String, Object> Mob)。

查询所有手机信息：public List<Map<String, Object>> getAllMob()。

根据手机名称查询手机信息：public List<Map<String, Object>> getMobByName(String MobName)。

手机修改：public void updateMob(Map<String, Object> Mob。

根据手机ID进行删除：public void deleteMobById(String id)。

4. Servlet 相关类

根据需要自行定义，如对应增、删、改、查功能的 Servlet。

5. JSP 页面

素材中已提供部分所需的静态页面，可以改成 jsp 页面。

提示：素材中涉及了 HTML 中 frame 框架的内容。

第八章 后台操作.pptx

项目源代码.zip

习 题 答 案

第二章

一、选择题

1. BC	2. D	3. BD	4. A	5. C
6. C	7. C	8. AD	9. D	10. BCD
11.	12. A	13. D	14. B	

第三章

一、选择题

1. D	2. AD	3. D	4. BD	5. B
6. D	7. A	8. D	9. C	10. D
11. B	12. D	13. D		

第四章

一、选择题

1. D	2. B	3. B	4. A	5. A
6. A	7. C	8. B	9. A	10. A
11. B	12. A	13. A		

第五章

一、选择题

1. B	2. B	3. B	4. C	5. A

第六章

一、选择题

1. D	2. D	3. B	4. A	5. BD
6. BD	7. A	8. C	9. D	10. D
11. C				

第七章

一、选择题

1. E   2. C   3. D   4. D   5. A
6. A   7. B   8. C   9. C   10. B
11. B   12. C   13. D   14. B   15. B
16. B   17. B   18. C   19. A

参 考 文 献

[1] 张爱玲. Java Web 项目实战教程[M]. 北京：机械工业出版社，2015.
[2] 宁云智，刘志成. JSP 程序设计案例教程[M]. 北京：高等教育出版社，2015.
[3] 王国辉. Java Web 开发实践宝典[M]. 北京：高等教育出版社，2010.
[4] 郑睿. JSP Web 应用程序设计[M]. 北京：高等教育出版社，2010.
[5] 刘素芳. JSP 动态网站开发案例教程[M]. 北京：机械工业出版社，2012.
[6] 明日科技. JSP 技术开发大全[M]. 北京：人民邮电出版社，2007.
[7] 冯艳玲. 中小型 Web 项目开发实战[M]. 北京：人民邮电出版社，2013.
[8] 杨光，伍连云. Java Web 实践开发完全学习手册[M]. 北京：人民邮电出版社，2014.
[9] 陈磊，徐受蓉. JSP 设计与开发[M]. 2 版. 北京：北京理工大学出版社，2016.
[10] 刘俊亮，王清华. JSP Web 设计与开发[M]. 北京：北京理工大学出版社，2011.